GHOST DETECTIVE

ADVENTURES OF A PARAPSYCHOLOGIST

By

DR. ANDREW NICHOLS

COSMIC PANTHEON PRESS

ISBN: 978-0-9834369-0-4

COSMIC PANTHEON PRESS

PUBLISHED BY COSMIC PANTHEON PRESS
www.cosmicpantheon.com

PROUDLY PRINTED IN THE USA!

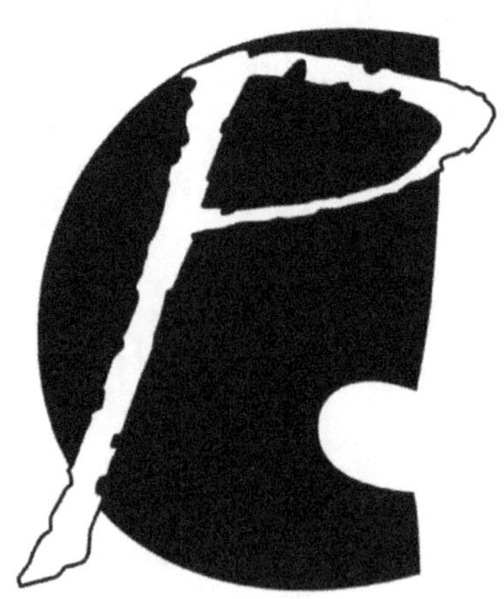

THIS BOOK IS PART OF OUR PSI FIELD RESEARCH SERIES
Other books in the series:

- Ultimate Ghost Hunter by Vince Wilson
- Aliens Above, Ghosts Below by Dr. Barry Taff
- Parapsychology: Frontier Science Of The Mind by JB Rhine

FOREWORD

BY
PROFESSOR ALVIN PROFFIT

I am honored to have the opportunity to introduce this book, Ghost Detective: Adventures of a Parapsychologist, and it author, Dr. Andrew Nichols to you. Ghost Detective skillfully reflects the art and science of ghost hunting to the reader in a unique way. The book chronicles the life's work in parapsychology of the man who has become known as the ghost detective (Andy to his friends).

With a charming blend of intelligence and humor Dr. Nichols recounts some of his favorite supernatural stakeouts. With refreshing honesty our supernatural super-sleuth admits that he does not always get his man, or woman, or whatever it is that he is chasing. Quite to the contrary, there have been times when his spectral quarry almost got him.

I first met Dr. Andrew Nichols when he was the guest star on Speaking of Strange, a popular paranormal radio show that originates in Asheville, North Carolina. The show's host, Joshua Warren, had invited me to be an in-studio guest adding local flavor. During one of the commercial breaks Joshua asked if I wanted to conduct the interview with Andy for a few segments. I gladly accepted. I think I have been interviewing Andy ever since.

Since that time Andy has proven to be a good friend and mentor to me in my study of the psychical sciences. Given that serious students of parapsychology have few opportunities to learn the science, and art, of psychic research though formal training, I have been blessed to have him help guide me in my quest to better understand and research the paranormal.

At the time I met Andy the first edition of Ghost Detective was hot off the press and I found it to be a fascinating read. Since that time it has become even more fascinating to me. As I got to know the man behind the words his stories took on a new life. His honesty regarding what it means to be a parapsychologist became more meaningful to me as I began to try and help those who were haunted.

Andy is a psychologist, professor, scientist, and researcher in his "normal" life. Yet the analogy of being a detective, that the media has chosen to describe his work as a parapsychologist, is quite appropriate. The similarities between being a detective and scientist are personified in Andy's work. As a scientist/psychologist he works to advance knowledge of the human mind where there is yet much work to be done. As a ghost detective/parapsychologist he diligently works to solve mysteries that offer few, if any tangible clues. Both are mysteries that have existed since the beginning of human history yet have never been fully or satisfactorily explained.

Even when paranormal clues are found they often contradict proper definitions of logic and run counter to the established wisdom of scientific rationality. At times these clues are as unbelievable to the scientist as they are to his or her skeptics. When confronted with such paradoxes the difference between the true scientist and the debunker is that the scientist keeps searching for answers while the debunker announces, at times arrogantly, that the search is over and was foolish to begin with. It is a good thing that the likes of Galileo and Copernicus didn't agree with their detractors that the search was foolish.

While remaining committed to the importance of scientific methodology Andy understands the necessity of being open minded while seeking answers to an unseen world that is reminiscent of the work of Louis Pasteur. Dr. Pasteur's theories regarding germs angered men of great intellect and learning who criticized and mocked his belief that something unseen could impact the observable and predictable world of medical science. While medical science now fully accepts the role of germs this kind of thinking and misguided logic sadly continues to prevail in many scientific circles. William James, J.B. Rhine, Carl Jung and Thomas Alva Edison also experienced this kind of professional wrath. Scientists expanding the frontiers of established knowledge are too often misunderstood, and are rarely appreciated for the work they do.

As a result it is imperative that a parapsychologist be his or her own best critic. Andy has taught me that when you offer ideas that run counter to the conventional thinking of science you need to be your work's strongest filter. In a world that affords parapsychologists little, if any leeway, offering natural and explainable events as evidence of the supernatural can lead to a loss of what little respect the parapsychologist may have had. Accordingly any mistake, or fraud, of a single parapsychologist hurts the entire profession.

Regardless of the degree of care the scientist takes with their research the constant devaluing of their efforts is not easy to endure. Years of having your work criticized, while enduring questions regarding your integrity, will take its toll on anyone. It will test the limits of commitment to your work. It helps explain why there are so few true academicians and scientists seeking answers to paranormal events. Much too often the work is left to unqualified charlatans and publicity seekers who further erode any sense of respectability for the psychical sciences.

Yet this has not stopped Andy in his quest to understand and explain paranormal events. It has never stopped the ghost detective from risking his professional reputation while working to solve the questions that haunt people worldwide. To paraphrase singer/songwriter Bob Seeger, Andy continues to work on mysteries, at times without any clues, and is always trying to figure out how the night moves.

As the ghost detective Andy neither sets out to validate or debunk any case he takes. He simply looks for the explanation and the truth. He has great respect for all people yet feels a sense of necessity to be completely honest with them. In return he expects them be honest with him, and themselves. An example of this can be found within the book in a case of a woman who pleas with Andy to help rid her home of her of a ghost that has developed a fondness to stealing food from her refrigerator. At the same time she reports that her young son, who naturally would never lie, is seemingly gaining weight faster than a mosquito at a Jack the Ripper crime scene. Your first name does not have to be Sherlock to figure this one out, but Andy's telling of the story is marvelous.

Another of the cases that Andy relates in Ghost Detective is driven by the actual transcript of an initial interview with an "afflicted" family. The transcript tells the story itself and is unintentionally funny without any editorial comment. Yet when Andy offers his comments you can almost see the look on his face. It involves a family that has summoned the ghost detective to their home because they are suffering from a variety of supernatural happenings. During his initial interview with them they nearly set a new record for the most paranormal events ever reported in a single interview. Even with this valiant effort, and probably to the family's dismay, the record is still held by George and Kathy Lutz's tales of woe and misery as told to Jay Anson in the now infamous Amityville Horror: A True Story. It is impossible to read the transcripts of Andy's conversation with the family and not smile. Of all the things I admire about Andy, high on the list is his ability to stay civil in such situations. His desire to help the family was sincere, but it became evident that they needed the services of someone other than a ghost detective. I just wish I could have been there just to see the expression on Andy's face as the family spun their tales of terror.

This story represents the fact that serious parapsychologists not only endure the slings and arrows from many academic and scientific directions they are often subjected to absolute lunacy, if not fraud. I am particularly fond of this case because it is eerily similar to the first case I ever consulted Andy on. It involved a woman whose haunting initially seemed mundane, yet disturbing. It quickly transformed itself into a case of "blood sucking ghouls" trying to kill the woman and her adult daughter. In short order the woman became more concerned with the fact that there may not be enough room for all three TAPS vans to park in front of her home and she needed new appliances to replace the older ones that had been ruined by "high EMFs" that the power company could not explain.

I am a bit embarrassed to admit it, but I called and discussed this case with Andy and he listened patiently to me. Without referencing the obvious he simply stated that, for more than one reason he drew the line at working on cases involving blood sucking ghouls. I am amazed he still consults with me. I just guessed that is just what good detectives do. When I related to the woman that I felt her problems may not be supernatural in nature she became quite agitated, told me I didn't understand and asked me for the number of someone who was good enough to be on TV.

There are so many fascinating cases in Ghost Detective that are in retrospect equally amusing, but legitimately terrifying to the victims. One such case involved an attractive single woman whose was terrified to learn that when she was not home her phone was being answered by what seemed to be a heavy breathing, snarling demon like creature.

After being told this by her friends she decided to try it for herself. Much to her amazement she called her number while away from home and was terrified to experience just what her friends had told her. Imagine the horror of hearing your own phone answered in this manner. It must have been terrifying for the woman to go home after that. Yet she kept her wits and had the good presence of mind to make one more call, to the ghost detective.

As you would expect the ghost detective was quick to respond and before you could summon forth Phillip Marlow on a Ouija Board Andy cased the joint and cracked this case of the demonic phone answering service by just the simplest means of gumshoeary. Let him tell you how.

Other cases Andy discusses were not as simple to solve and are without any trace of humor. While Andy can see the humor in many situations, which is a good trait for a parapsychologist to have, he is extremely serious with his work. As a psychologist, his chosen trade, he fully understand the potential toll that a haunting, regardless of its source can have on a person. There are people who hurt from supernatural occurrences in their life and sometimes outside help in an unbelieving world is their last vestige of hope. Even if they are delusional and their problems are of their own making they nonetheless need help. Professional help. Andy is an example of a man who can, and will, help those who are in the most need.

Andy is a trained scientist, a consulting psychologist and college professor. He has worked, at time unceremoniously, with some of the world's most perplexing paranormal cases. He has the distinction of being published in both the academic and commercial arenas, and was the recipient of the largest grant ever awarded to a parapsychologist for the study of poltergeist disturbances. He has traveled the world over in his role as the ghost detective. He appears often on national television and radio, and is a headline presenter at major paranormal conferences. Since 1993, He has been the Director of the American Institute of Parapsychology. He is the real deal.

Yet, some of his ideas are radical even for his fellow parapsychologists. At times he finds that his theories place him outside any circle you could draw. While it is common to say that we need people to think outside the box, there are times Andy can't even see the box from where he is standing. And in this business this is just what we need. I have enjoyed a long standing history of joining with Loyd Auerback, another of the absolute superstars "in the business", in debating Andy about the realities of the unreal.

What Andy believes about the power of the human mind is as phenomenal as any ghost or poltergeist you could conjure up. As a professor of education I have studied the mystery of the (as of yet) unexplained portions of the brain we seemingly do not use. I think that Andy's theories may help explain this. Perhaps with time the ghost detective's research will help us better understand the puzzle of human consciousness.

It is clear that even though I consider myself a friend of Andy I am also a fan. I make no apologies for this. Having him as a colleague in the fascinating world of parapsychology has meant more to me than these words can convey.

The fact that Andy is my friend is not why I recommend this book to you. I recommend it because I believe that Dr. Andrew Nichols is truly the ghost detective. He combines common sense and intellect in a no-holds barred attempt to find an elusive truth in a most contrary and complex science. Case closed.

Prof. Alvin C. Proffit, Ed.D.
Appalachian State University
Boone, North Carolina – March, 2011

TABLE OF CONTENTS

FOREWORD...3

Acknowledgments...9

I n t r o d u c t i o n...10

Chapter 1: Ghosts are People too!...22

Chapter 2: The Haunting Entity...38

Chapter 3: The Game's Afoot!...83

Chapter 4: So You Wanna Be a Ghost Detective?...90

Chapter 5: The Haunters and the Hunter...101

Chapter 6: Busting the Ghosts: A New Approach...131

Chapter 7: In the Grip of Evil...136

Chapter 8: The Terror that Comes in the Night...146

Chapter 9: Body Snatchers...159

Chapter 10: Into the Unknown...164

Chapter 11: An Exorcism Made to Order...182

Chapter 12: When Common Sense Falls Victim to Uncommon Events 209

About the Author...213

Acknowledgments

Anyone writing on a subject which has more than its fair share of skeptics must surround themselves with people who are open to unconventional viewpoints. I have been fortunate to find family and friends who were willing to discuss the paranormal with an open mind, and who were more than generous with their patience and advice.

First and foremost, I owe a debt of gratitude to all those who willingly shared their ghostly experiences with me, and allowed me to question them at length about the details and personal meanings derived from those experiences. It is my hope that their openness will encourage others to view their extraordinary experiences with greater conviction, and the realization that help for the haunted is available if they seek it.

I am grateful to my colleagues in the field of parapsychology who read various parts of the manuscript and provided valuable suggestions, especially Professor Loyd Auerbach, Dr. Barry Taff, Vince Wilson, and Professor Al Proffit of Appalachian State University.

Most of all, I am indebted to my family, who has encouraged me in this project from the beginning. My brother, Randy, my parents, Nora and Robert Jackson, my children, Jonathan, Jared and Brendan, who had to put up with me all these years, and never let up on their abundance of love and support, and especially my wife and best friend Jerilyn – thanks for loving me back.

Introduction

"Do you know how weird it is out there?"
Bill Murray, Ghostbusters II, Columbia Pictures

The poltergeist is in full haunting mode. Pounding noises emanate from the walls, the sound of rocks raining down on the tin roof of the old house. An overpoweringly evil *thing* threatens us with death. Hurled by no human hand, a two-by-four rockets up the stairs, narrowly missing my head and striking the wall behind, falling to the floor with a clatter. Everyone screams; no one louder than me. Racing to my car to put as much distance as I can between that haunt and this hunter, I gasp, *My God! There really are ghosts.*

The idea for this book began the night I beat a hasty retreat from the Lusk House, a Victorian mansion in Florida's Ocala National Forest. That was 1977. I'd been chasing these mischief-makers for two years by then but never expected to catch one. How do you get hold of a delusion? When we lit the seance candles and invited the phantom of the mansion to join us I never dreamt we'd encounter anything; I was ready to tackle the Devil himself... After all, ghosts were just psychologically generated mechanisms: aberrations, hallucinations, illusions unsupported by fact and only rarely corroborated by others--people invariably carried away by the moment. Or else they were frauds. Whatever they were, they weren't real. I was wrong! Only someone who has never lived through a genuine haunting--come face-to-face with terrifying phenomena themselves--can find comfort in such thoughts.

Observed among the wraiths that followed were solid, fully formed flesh and blood entities; faintly luminous, smoke-like misty forms; and glowing globes or balls of light. Nearly imperceptible visitors left indentations when they walked and where they rested. The sound of their footsteps fell in hallways and stairwells. They moaned, wailed, howled, cried, laughed, sang, shouted, and talked in muffled voices. Unseen musicians played for large crowds of noisy, appreciative but invisible folks. Some smelled of rotten eggs; some of shaving lotion, or perfume (lilac is a popular scent on the "Other Side"). Others had a ham hock and bean aroma about them. A few were as cold as the graves they'd left; still fewer, happily, warmed things up with a flash of spontaneous combustion.

One was an infant who whimpered all night. Another, the essence of an eleven-year old illiterate murder victim, managed to communicate in drawings. A mother paid an earthly visit to her alcoholic son to scold him bitterly for falling off the wagon. A daughter called very long distance (from eternity) to alert her mom to the dangers of a cancerous growth. A suicide, still grumbling about "That damn Roosevelt and his New Deal," lived on in his 33 room house. There was a jogger whose demise didn't put a crimp in his morning routine. A truck driver who hovered aloft in his good ol' LA-Z-Boy recliner. Tavern patrons who refused to abide by man or Nature's curfew: they just didn't know when to say when. And these were the "friendlies."

On the dark side were the voluptuous old hags who came-a-calling at night. The curse inflicted upon an explorer of Native American burial sites; and another against a family of ghoulish locals for what may have been the same offense: plundering the dead. Folks were pinched, punched, scratched, bitten, smothered, and sat on. There were sightings of human and animal abominations: misshapen or incomplete forms; hairy, bristle-like creatures that clawed their human prey; ill-defined menacing shapes; and four legged dog-like critters that traveled between visible and invisible with apparent ease. These little "doggies" stared out of glowing red eyes. Take my word for it, you wouldn't want one of them to follow your kids home.

Knight of the Living Dead

"Who ya gonna call?" *Ghostbusters!* said the millions who lined up in the mid-eighties to see the smash hit movie from Columbia Pictures. But who ya gonna call in real life? According to Vanessa Gilchrest, UPI, "If you're having trouble with a ghostly, unexplainable chill permeating the family TV room, or an unruly poltergeist making sleep next to impossible ... you need someone to chase out that ectoplasmic nightmare. Who to call? ...Dr. Andrew Nichols." But I'm not *Ghostbusters.* I don't always get the ghost and I never get the girl.

So what's a Ghost Detective? Take away the ghostbusters' proton pack, jump suit, and fifty-million dollar budget; make him sit around dank old houses waiting for something--*anything*-- to happen, and what you have left is me. Thankfully, my adversaries have been less spectacular than Gozer of Babylonia.

This is my story and the story of the haunted houses and haunted people I've investigated during the past two decades. *The Ghost Detective* is an honest and factual account of the experiences I shared with my clients--the victims of these sobering disturbances. It was taken verbatim from tape recordings, from hand-written notes, and personal recollections. It is, to the best of my ability to relate, the faithful depiction of my adventures. As founding President and Research Director of the Florida Society for Parapsychological Research, I have served as master of ceremonies, organizer of research projects, and procurer of guest speakers. In 1993, I formed the American Institute of Parapsychology. The AIP is a counseling and educational organization dedicated to the advancement of parapsychology as a profession, and to serve as a resource center for those wishing to explore the meaning and transformative value of psychic experiences.

Recently, in the middle of "Sweeps Week," a frantic time for program directors as they press to garner more than their share of the media's bread and butter--the ratings, MSNBC aired a program called *True Believers.* Me and some of my competitors were among those invited to display our talents. As a professional parapsychologist, I frequently appear on the tube. The host introduces my competition to the TV audience. I've never seen them in action, but I'm told they locate the spirit's "cold spot," encircle it, then move as one body slowly toward the front door. After which, the door is thrown-open and the invading spirit thrown-out. According to one of their members, "What we do is ... surround the entity using mental thought patterns-and ship it off." "How do you get rid of 'em," she asks their leader? (I guess she means if they refuse to ship-off.) "Well, we say prayers and just hope that they'll leave," her reply.

A professor colleague asks me, "With all the weird characters you must run into, how can you take any of this stuff seriously after listening to them?" "It isn't easy," I concede.

In my eagerness to help them, I'm sometimes exposed to crackpots and dingalings. Peculiar to the subject of things metaphysical are the mildly to severely afflicted people out there looking for a platform to voice their maladjusted opinions and paranoid delusions. Concepts like apparitions, telepathy, psychokinesis, and especially doctrines that advocate the black arts are appealing to the borderline personality. By appearing to bend the laws of cause and effect, these notions give a kind of legitimacy to their own cockeyed view of the world. The tales they tell range between amusing and absurd.

Finishing up a parapsychology lecture at the University of South Florida, I was approached by a young man who greeted me with, "Here I am! Got your message."

"Pardon me?"

"I heard your lecture." Putting his finger to the tip of my nose, he explained: "I knew you wanted to talk to me, so here I am. What did you want to tell me-about Madonna, I mean?"

"I'm sorry, but I don't think I have anything to tell you." I figured the guy was a few bubbles short, but since I had been paid to appear at the university, I ought to at least hear him out. It seemed he was having a relationship with Madonna-if only in his mind:

“She's crazy about me!”

“Who is?”

“Madonna!”

“The singer?”

“Yeah!”

“How do you know?”

"She couldn't sing to me the way she does if she didn't love me."

"You mean, sing to you in person?"

"Oh, no! I could never be with her that way. I mean on tapes and television. It's the way she looks at me."

Psychologists call these delusions symptoms of "erotomania." I listened attentively while this man described his powers of ESP. Example: He always knew when his favorite football team, the Dallas Cowboys, were going to win. How? "It's simple," he told me. "It's the uniforms. When they wear white jerseys, they win. When they wear blue ones, they lose. I wrote them about it but they never wrote back..." His intuitive abilities were-well ... different. But what did they have to do with me and a girl named Madonna? That was simple, too. It seems that shortly after hearing me speak on a local radio show the station played one of his dream girl's recordings. Convinced there was a connection between my talk that afternoon (which happened to mention telepathy) and Madonna's rendition, my visitor made a beeline for me to find out what it was.

I'm grateful that not many of them show up at my door. More commit their ramblings to the telephone and the written word. Sometimes they insist on blaming spirits rather than facing the truth. Lake the indignant woman who said a ghost was "stealing leftovers right-out of the ice box." "It's hard enough," she grumbled, "to keep little Bubba (her obese thirteen-year-old) out of there .....he's on a strict diet, ya know. And food's disappearing right and left!" Had she considered the possibility that little Bubba was sneaking a snack or two? "I thought of that. But when I asked him, he said, 'No way, José! And one thing I know for sure, my boy never lies."

As far as I know, ghosts do not raid refrigerators--at least not as often as chunky kids. Overweight individuals do lie about their eating habits; even pilfer leftovers. So before committing myself to a search for an ethereal glutton, I suggested that Bubba might be *unconsciously* nibbling when no one was looking. "No! It's a ghost!" she persisted. "My boy don't lie. What are you going to do about it?" "Put a lock on it lady," I said, meaning the ice box, of course.

A middle aged man wrote, "I am a proven case of reincarnation and an alien on a secret mission." One caller claimed to have raised the dead, while having remained, "Not the least bit eccentric." He added, "If I am The Christ, I haven't noticed it yet." Another guy enjoys the stories Mickey Mouse and Donald Duck tell him; not to ignore Warner Brothers, he hears from Bugs Bunny, too. A retiree living in Florida has a post-life tape of Richard Nixon speaking, as he tells it, "Presidentially" and yet, at the same time, "Down to earth." According to my correspondent, Mr. Nixon recorded it several years after he passed over to wherever it is old Conservatives go. "I want twenty-five dollars a copy," he says. "People wouldn't pay twenty-five dollars to hear Nixon when he was alive," I reply. A woman told me a spirit suggested she write: "I'm 100% sure I can prove that I have communicated with the spirits of the famous," which she lists (seventy of them) on

the reverse side. “It might be wondered just why they tell me things,” she continues. "It's because I have the mind
to understand what they tell me ... more or less, that is."

And there's the plea I got for relief from “psychic torture”: “Years ago I wrote letters to Elvis. After he died he came through on a psychic channel to me and caused me, for five years until now, this terrible bodily pain.” The alleged persecution was confirmed in a postscript: “All the above is true. I am her daughter, age thirty-nine. I have seen firsthand the extreme torture my mother has undergone since she wrote to Elvis.” I can't help wondering if this poor woman's in hiding now that the deceased Elvis has been spotted in so many places.

Screening prospective clients is an absolute necessity. Over the years I've learned to master this sometimes disagreeable chore. I exchange letters and e-mails with them; speak with them on the telephone; and sometimes drop by for a chat before an “official” (when I do more than just talk) house call. It's a step-by-step procedure that screens-out the "flakes": the ones who, although probably not certifiable are, to be polite, more than a bit *strange.* The majority of my clients are ordinary, level-headed folks who are misinterpreting some natural event as a supernatural one. Most are what we like to think of as normal. Some are not. It's often what they write, and sometimes the way they write it: tiny words that wind their way around the perimeter of umpteen pages, interspersed with a good many racial and religious slurs directed at the human race in general; phone callers who mention that they're not sure, but they might be "physic," or even "psychiatric." I never argue with them.

Many insist on telling me about fantastic creatures, invisible malevolent forces, and sinister groups who are bent on ending the American way of life. Some wear Reynolds Wrap inserts under their hats to protect them from "The Rays." Others carry a heavy crucifix, or adorn themselves with garlic to ward-off vampires. There's a generous supply of otherwise average men and women who are plagued by accidents, diseases and just plain bad luck. Many believe they've been cursed, or that the forces of darkness are out to get them. Of course, none of these gripes have anything to do with psychic matters, but they've exhausted their supply of listeners by the time they get to me, and now it's my turn.

It's not that I'm unsympathetic towards them. I do answer their questions to the best of my ability. If they're not kidding, however, they need a lot more help than I can give them. Psychotherapists know that buried within the deep recesses of even the balanced mind lurk the vestiges of unrestrained fantasy: primordial man hiding in the shadows of prehistoric magic from demonic creatures, ghouls and ghosts. Sometimes they surface to play havoc. A few, it seems, break into the conscious sphere clutching my name, address and phone number. After all these years, I guess I've heard from as many "psycho-ceramics" (my term for crackpots) and just plain mixed up people as anyone. But that doesn't stop me from doing it. Like the guy who joined the circus and was admonished by friends to stop shoveling elephant chips said, "What, and give up show business?" I refuse to give up.

This brings me to an area of concern I share with some of my parapsychologist colleagues. Our difficulty centers on the problem of deciding which cases to take and which ones to avoid, because they're just too goofy to bother with. In the early years, the late seventies and early eighties, I was eager to get as much exposure to hauntings as possible; I took nearly every case that came along. Once I had screened-out the obvious flakes and those who had nothing going on in their lives even remotely resembling psychic phenomena, I felt a moral obligation to help the rest: those trying their best to maintain a normal lifestyle while coping with the most frightening kinds of disturbances.

I always believed that merely the act of counseling these beleaguered folks would afford them a measure of relief. Other parapsychologists have individual ideas of what constitutes a "crackpot case," and the extent to which our counseling should progress.

As a rule, the more bizarre the claim the less most parapsychologists are willing to get involved. Still, it's important sometimes just to listen to what they have to say. I investigated such a case in 1988, precipitated by what my caller said was a female Neptunian (that's right, a being from the planet Neptune). My curiosity was piqued. The people were challenging; the culprit turned-out to be--surprise!--not a space creature, but, as I saw it, a psychological projection enhanced by someone's over active imagination. In recent years, colleagues at parapsychological research institutes (such as the Rhine Research Center) have referred a number of investigations to me because, as they put it, "We're tired of listening to kooks." "I can understand why psychiatrists go daffy when they're dealing with these things day after day," they say. "It's really unfair on our part to be impatient and intolerant of them, but when they start saying they just came down from Neptune and they're on a secret mission ...C'mon!"

They marvel at my patience (they're too kind to call it gullibility). I remind them that if it weren't for the "kooks" we'd have no cases at all. Except for those that have a natural explanation, none of the stories we hear could be considered "normal" by anybody's standards. Those who have extraordinary tales to tell should be excused if they appear the oddball when they tell them. On the other hand, sometimes my colleagues' reluctance to get involved is justified. It doesn't happen often, but if I haven't done a proper job of screening I may find myself the guest of some really far-out people.

It may appear otherwise to the casual observer (as well as to my colleagues) but only a small percentage of the reports I receive openly suggests mental aberration or outright fraud. Many do turn out to be misidentified natural events, but I discover it only during an on the scene investigation.

Take the following for example: In the seventies, shortly after I took the psychic investigator oath, I investigated a phone answering spook. A divorcee living alone with her Labrador Retriever, Dusky, declared that someone was answering her telephone while she was out. Friends said that when they called, someone would pick up the phone, but instead of "Hello" they'd hear deep breathing.

Now many people, especially women, find themselves on the receiving end of "heavy breather calls," but few initiate them. Thinking the equipment was on the blink she called Ma Bell. There was nothing on the line or in her unit that could account for the sounds. "I tried myself one day," she said. "I called my number and sure enough 'It' answered. I swear, to me it sounded like The Beast, snarling and panting. It scared me! Can you imagine, here I am all alone in this house with God knows what?"

As a fledgling Ghost Detective in the early days, I made all the house calls I could, and did them gladly. I was there for about thirty minutes when the phone rang. I put my hand up to stop her from answering. It rang twice more, when out of the kitchen came Dusky scurrying along to the alcove that housed the telephone. It rang again. He looked dolefully at his mistress, turned his head toward us then back to the phone, clenched the receiver in his mouth and lifted it off its cradle. I suppose it was my fault. Had I questioned her more thoroughly I would have learned that the receiver was always found dangling in mid-air. I'd have saved a trip. Animals don't appreciate high pitched noises, and this particular instrument rang at the very top of the octave. Dusky spent much of his life alone in the house. His mistress, a recent divorcee, was young, attractive and popular. Her telephone was, figuratively speaking, ringing-off the hook. The dog had merely learned to stop its annoying peal.

Parapsychology and Me

The author of this book is a professor of psychology who has spent what seems like several lifetimes investigating paranormal phenomena. In my youth, an uncommon interest in abnormal psychology led me to psychical research. By the time I was twenty-five, I had read every piece of literature I could find on ESP, ghosts and hauntings. After exhausting the public library I began to accumulate a sizable one of my own, including works written by little known authors. An analytical person by temperament and profession, I have the patience to spend days in the libraries and record halls; then more time spent writing-up the tedious reviews and opinions. I can also be found conducting preliminary interviews for a psychic investigation, researching the history of the house, setting up the cameras and other equipment for a ghost vigil. It's my bailiwick. Another perk of my profession is the numerous appearances on TV programs featuring paranormal topics including *Unsolved Mysteries* and *Sightings.* Both programs featured my investigations of numerous haunted properties. I've enjoyed most of the media attention.

My *thing* is to gather and interpret empirical data on hauntings. I think I do that as well, or better, than anyone else. I know that many of my cases would be hard-pressed to stand up to scientific scrutiny; but then I don't investigate hauntings to establish proof of their reality. What I do is not a laboratory science, it's a "street science." I operate without benefit of university or other official sanction in a world where reason is often up for grabs. Yet I staunchly defend my scientific right to tinker there. Why? Because nobody does it better. Moreover, because only a handful of scientists will even try.

Observing ghosts firsthand is a treat reserved for the few; therefore I must usually rely on the testimony of those unique individuals who report them. In this respect I bear more resemblance to a folklorist than to a laboratory-bound parapsychologist: My methods differ from those of some researchers who document and measure hauntings. Proving that something bizarre occurred, then establishing its parameters is an important function; but learning what the observers *think* occurred and how they react to the ordeal may be even more important. Naturally, separating fact from folklore is part of my job. Still, it doesn't interest me half as much as giving haunted house victims comfort, helpful advice, and when possible, relief from the offending intruder, whatever its source. As a consequence, before agreeing to take a case I don't require that the phenomena be scientifically proven or, for that matter, even scientifically possible. If I did, I'd have precious little to investigate.

I accept the accounts of my clients for what they are (at least at the outset), but I'm far from credulous. As any detective will tell you, relying entirely on the testimony of an eyewitness is not always the path to the truth.

Psychic Philanthropy

Why do I do it? Why study events that are, in the main, totally explicable? Why waste a lifetime risking your reputation on a subject that no one but the zanies believe in? I do it because I enjoy it. Some people play golf. I don't play golf. Well, I do, but the other people I'm playing with don't enjoy it. Seriously, why do I hunt ghosts? Certainly not for the money; I've yet to earn a full time living at it. And whose fault is that? Mine. Except for media appearances, I rarely charge for what I do. But think about it. How could I charge for services like mine: by the hour, by the number I "ship off?" Anyone performing a legitimate service is entitled to charge a fair fee. Yet, when you consider the days and weeks and sometimes months parapsychologists devote to a single case, how could anybody but the wealthy afford me? I'd eliminate some of my best cases if I charged.

Haunted houses attract me for a number of reasons, not the least of which is curiosity. High on the list, too, is a genuine desire to help people understand and cope with their problems. It may be a cliché, but material wealth isn't everything. In a society that seems to exalt the philosophy of "What have you done for me lately?" little is done for anyone without the expectation of reward. There is another kind of compensation that has nothing to do with cash. In addition to our devotion to family and friends, each of us has an inner need to make some contribution of himself to his fellow man. He may do that on a personal level through social service clubs; by working in the community with children, senior citizens, and the ill; or by volunteering for a myriad of other worthwhile projects. Part of my contribution is the patient understanding and the expertise I give to the victims of hauntings. The reward is immeasurable.

Because I refuse to violate the confidence of those who have graciously invited me into their homes and lives, I have used fictitious names, and, as a further precaution against giving away their identities, false locations when telling my client's stories. Authors who refuse to divulge the source of their material are not always taken seriously. Even *FATE* magazine (as stated on its cover) publishes only "True Reports of the Strange & Unknown." Before an article can be accepted the publisher insists on names as well as statements from those directly involved. They're kept on file in case proof of authenticity is requested by so-called responsible researchers. Publications have legal as well as moral obligations to their readers.

Although some had no qualms about being identified, the majority of those I contacted did; they voiced strong opposition to the idea. The last thing I want for them is notoriety: I've turned down cases where the prospective client hedged his or her invitation on my willingness to bring along a member of the media. There's no doubt that genuine phenomena manifest in homes where the residents look for publicity. Outside influences, however, are counterproductive. Anyone encouraging such attention draws suspicion to their motive for crying "ghost."

For similar reasons I shy away from ambulance chasing--or a better term might be "hearse chasing"--newspaper accounts, or stories on the evening news. Publicity received before my arrival tends to add confusion and obscure the real cause of the haunting. Human frailty leads some to embellish facts; to prolong their moment in the spotlight by "fudging" a little (often unconsciously). Fortunately, I have ample cases without them. Yet it's safe to say that I've missed-out on some evidential ones by shunning them. My unwillingness to give names and addresses results in the loss of credibility: you're no more likely to see my adventures in *FATE Magazine* than you are to find them in *Scientific American.* It's not a cop-out. It's just that I intend to continue in this business until I'm either a member of senility's "La La Land," or a ghost myself. I'd like for there to be haunts available for me to hunt. That's not going to happen if prospective clients lose faith in my ability to keep their secrets. They share their tales of woe with me because they know they won't find their life's story, or the faces of their children staring back at them from behind the checkout counter. Anyone who has had the misfortune to live in a haunted house or who has studied them in the field will immediately recognize the authenticity of the phenomena and the spontaneous reactions of the people in the stories that follow...

Fact is a Poor Story Teller

"I can't get over it," a magazine writer once said to me. "You sit there and talk about these things as though they really exist; but do you believe in them yourself?" I've had to ponder that one many times. It's a tough question because there's no stock response, no simple reply. Most people in this hurry up world, especially those in the media, are chronically short of time and patience. The last thing they want to listen to is comprehensive mumbo jumbo. But brevity, no matter how much appreciated, is sometimes mistaken for superficiality. Ghosts and hauntings are a knotty study; much more complex than the antics of Murray, Aykroyd, and Ramis would have you believe, yet far less serious than Blatty, King, and Straub would have you imagine.

Simply stated, I believe in ghosts. I believe, not because seeing is believing:

we who investigate are seldom privileged to witness them up close. Except for a few brief encounters of my own, I have only the experiences of my clients to evaluate. I accept their reality because of the accumulated evidence of the things they do, and the effect they have on their victims: a theme that begs to be set down, and one elaborated upon throughout this work.

A danger exists in drawing judgments from a limited number of cases. This modest study hardly examines enough material to consider it conclusive. Yet it does parallel and confirm the work of dozens of other psychic investigators over the past century. As a result, I've come to believe that ghosts really do exist and hauntings do occur; that those who report them are not always misidentifying natural events; they're not all crazy, or perpetrating hoaxes, but have experienced real and not infrequently terrifying phenomena. I did not recount my adventures, however, to promote any particular theory about the supernatural. For all I know there may not even be such a realm. I undertook this task, primarily, to inform the reader what seemed abundantly clear to me from the beginning: that whatever their source, the phenomena that manifest during most hauntings are real. What's more, under the right circumstances, it is entirely possible to rid the premises of these annoying and unwanted displays of energy. I cannot pretend that my work will make a substantial contribution to the science of parapsychology. It certainly isn't a textbook on it. Much of what I have said is well known to students of the subject. As for the rest, the reader may draw different conclusions from the spontaneous cases I've researched, but all the facts are there. My mind is open to other interpretations; I'd love to hear yours.

The specific area of Spectrology--the study of ghostly entities and the hauntings they inflict--is only a part, albeit an integral one of psychical research. I've made no attempt to cover other kinds of weird sightings, even though there does appear to be an underlying connection between them. Since there's a lifetime of work left for me to do in my own field, I'm perfectly happy to concentrate my efforts on apparitions, ghosts and hauntings. I'll leave the study of religious visions, folklore creatures, extraterrestrials, men in black, and assorted ghouls and monsters to others.

It was no easy task to decide which of my adventures to relive in their entirety. Whether they were concluded in one evening or over a span of years, they've all been uniquely interesting. Hauntings are a people problem. *The Ghost Detective* is a tribute to the intrepid forbearance of my clients; they all deserve to be heard.

Literature abounds in ghost stories--enough to fill several libraries. Most follow a predictable path. Lovers of the arcane, like detective story enthusiasts, take pride in guessing the course they'll take after only a couple of pages. I think you'll find mine have a slightly different twist. Unfortunately, my cases don't all unfold in characteristic fashion. The bulk are composed of anecdotal material: little plot, no climax-Just a small number of brief encounters. There are few good story lines in real hauntings. As Somerset Maugham put it, "Fact is a poor story-teller." Those who write thoughtful studies on the subject usually give an accurate portrayal, one that's free of exaggeration. Their purpose is to shed light on the unknown. As a consequence, their books never become best sellers, nor are movies or TV series made from them. Controversy piques the interest of men. Fact, although stranger than fiction, often bores them to death.

Who should read *The Ghost Detective*? To the incurably incredulous I say, don't waste your time. If you're looking for proof that the things described herein originate in the world of the supernatural, you won't find it: not here or anywhere else--for none exist. A reluctance to jump to supernatural conclusions is normal and healthy.

Several years ago, I witnessed it in the face of overwhelming evidence to the contrary. By sheer volume of reported effects this household was one of the most overtly haunted I have ever come across. The man of the house, who spent the better part of his life on the road was, however, blissfully unaware of it--that is until the following incident: "I just came in," his wife told us. "He was snoring-away on his recliner. Everything seemed okay, except when I looked again I saw him and his chair floating about this much (she indicated a good foot) off the rug. I couldn't help it ...I let out a scream and down he came, crashin' to the floor."

Our napping aerialist shook his head up and down. When I asked him what he thought about it, he said, "Yeah, there's been a lot of funny things going on. I guess I was off the floor some (he showed me where the fall had cracked the chair at its wooden base], but I still don't believe in no ghosts!" (America's Most Haunted.")

I share many of the doubts of my clients, plus a number of my own. They're uncertainties similar to those of the disbelieving husband--related to the origin of the disturbance and not to the disturbance itself. However reluctant he was to admit it, our client knew he and his chair had been airborne. What he was unwilling to accept was the possibility that a ghost had made it happen. There had to be a reasonable explanation for all the "funny" things going on, and it was my job to find it. And yet, isn't there always a reasonable explanation for seemingly supernatural occurrences? No one can nullify the laws of gravity. No one can just float off--not physically, no matter how comfortable they may be.

I cannot make an iron-clad case for the existence of the unseen forces or the unexplained happenings described in this work. Even if I could, you super-skeptics wouldn't believe in "no ghosts" anyway. And all you "true believers": You'd be disappointed, too, and for many of the same reasons.

Sir John Eccles, who won a Nobel Prize for medicine, has said that science cannot answer such fundamental questions as: Who am I? Why am I here? What happens after death? As we enter the twenty-first-century there are those who think these mysteries are beyond science. Yet I believe what is inexplicable today only awaits clarification and definition tomorrow. No doubt a reasonable explanation for ghosts and hauntings exists and will eventually be found. In the meantime, there are an awful lot of folks out there who can't wait for an indefinite tomorrow to get the help they need. In the following pages I attempt to explain how parapsychologists go about providing it.

Although I have speculated and philosophized more than I intended, I hope the reader will not find the work preaching or moralizing. It's not up me to tell *why* these things happen, I can only relate what took place and give some of the theories advanced to explain them; only God knows *why!* My goal is to present my experiences as accurately and honestly as I can. I leave the final judgment as to their meaning and value to the reader. Finally, the opinions and conclusions expressed are, unless otherwise noted, all mine.

Chapter 1: Ghosts are People too!

"All houses wherein men have lived and died are haunted houses."
(Longfellow, 1858)

CASE FILE: A PHANTOM IN WHITE

Some houses hold more than just a haunting trace of their former occupants. Some hold psychically imprinted energy patterns, images from the past that are played again and again for their rapt audiences. For reasons explained later, it isn't always possible to identify the initiator of these movie-like replays. Moreover, because they're only *reflections* of the past, there's seldom a motive (or, more accurately, one that we're able to recognize) for the things they do. In 1990, I studied a series of incidents in which the traumatic experiences of a terminally ill young man seemed to have created a long-lasting psychic memory. In this account, as in most of those chronicled in this book, the names of the participants have been disguised. That's the way the people involved wanted it. I assure the reader, the things that happened to these disquieted people remain unaltered.

Carl and Geri Kramer own a boarding ranch for riding horses in an affluent rural area of the Central Florida. Over the years they've built a successful business stabling and training their charges.

In March 1989, the Kramers purchased the land adjoining theirs, increasing their spread from ten to forty-three acres. The new property included a modern ranch house, which until just after my arrival was occupied by their twenty-four-year-old horse trainer Buck Shepard and his black Labrador retriever Thor. Two other hands, Kurt and Wes (also in their twenties), had cleared out several months before. It was just after the house was repainted and refurbished (in August of that year), and on the heels of Buck, Kurt and Wes moving in, that strange things began to happen.

During one of our phone conversations Mrs. Kramer told me, "The weekend the kids moved in they heard somebody come through the back door, walk across the kitchen floor ... which you could hear because it's linoleum ... and then go out the front door. They all heard it, but it was Kurt, in the front bedroom, who jumped up and looked outside. Nothing was there. After that they kept hearing doors opening and closing and finding the back door standing wide open." Geri went on, "Last October, just after Halloween, Buck was working in the barn when he caught a glimpse of a man in a white shirt go around the side of the house onto the deck. He went down there to check it out but couldn't find anybody. "Thor...Buck's Black Lab...very often he'll stare at the fireplace. The hair on the back of his neck stands up and he'll jump and bark at nothing. Once Buck saw him follow something down the steps with his eyes, but couldn't see anything himself. He sleeps with Buck and is almost always in sight of one of them or the other when things start to happen. "All the while they're living there they hear things at night...like something dropping on the glass-top coffee table in front of the fireplace. You know, just a bump in the night kind of a thing. They'd hear doors open and close every once in a while, but they always managed to ignore that stuff. ..play it down. "Wes was up one night watching TV after the other two had gone to bed, and he watched as one of the bedroom doors opened and shut twice. He just got up and went straight downstairs to bed. He left the TV, the lights, everything on and went to bed! "All the lights in the house turned on by themselves once; and another time Buck and Kurt were in the kitchen when the garbage disposal came on by itself. "Buck told me, 'You know, I think somebody came into the house last night.' And I said, 'Buck, you guys never lock your doors.' Well, they started locking their doors, but they still heard them open and shut at night."

In an event that occurred earlier, Buck and Wes were working down in the barn when they decided to come up for lunch: "They had just entered the house through the back door when something in a white shirt went from the fireplace through the house, maybe twenty feet in front of them. "As I am speaking I am getting goose bumps," she added. "It went, like, toward the front door and they were at the back door. And, right away, Buck thinking someone's in there, yells, 'Hey, what ya think you're doing?' "Every time they tried to talk about what they actually saw that day they could never pin it down; never come up with a definite description. Except that it's a guy in a white shirt. Buck finally admitted that the house is haunted, but that didn't keep any of them from living there ... at least not in the beginning. Maybe because it's rent free."

On June 16, I paid a visit to the ranch where Carl and Geri Kramer, their fifteen-year old son Jason, and Buck Shepard filled us in on the alleged haunting in person. Since the sudden departure of ranch hands Kurt and Wes, young Jason had been added to the list of earwitnesses. Geri explained: "Buck's often too nervous to stay alone, so Jason spends his nights in the house keeping him company." The teenager described his experiences to me: "No matter where you are in the house at the time, the sound of the door slamming is like a big boom! It's right there with you." "Why had Kurt and Wes left?" I wanted to know. "One went back to school, the other was working in Minnesota now," Buck replied, with a touch of humor in his voice.

Between my last conversation with Mrs. Kramer and my visit to the ranch, Buck had seen the vision again. He spotted it one evening on the deck, leaning over the railing: "It looked more like a person this time, except I didn't see any legs. We've never seen any legs, and that's what scares us," he said. "I thought it might be Jason because he had a white shirt on; so I came back in here (the living room), but he was sitting on the couch listening to the radio."

The last incident Buck was made to endure occurred only days before he was scheduled to move out. An unnatural light had disturbed his sleep. There, standing next to his bed was a glowing figure in white: "It wasn't like you saw arms or hair or anything. It was more like a white blob. I could only see its chest and it was white. "I don't believe in ghosts myself, to be quite honest with you," he said--reciting the time-honored talisman employed by the frightened--"but the experience made me thankful that they (the Kramers) have another place for me." (Shortly after my visit, Buck moved into a small farmhouse on the property.)

Carl Kramer offered a history of the place: as much as he knew. It was built by Buster and Olivia Niles in 1982. Buster, a pseudonym for a prominent entertainment figure, divorced his wife just before moving in. Mrs. Niles and her terminally ill son Frank lived there until Frank's death. They had been joined by her second husband in '86. "Mercifully," Carl went on, "the ailment took the boy's life in 1988. He was just twenty-five." Later that year a man and his wife bought the house, stayed there a short time, then suddenly left. It remained vacant until the Kramers moved in.

We were curious about the ramp up to the front door. Geri told us that in the beginning Frank was walking: "His parents must have known that his health was deteriorating and that he'd be confined to a wheel chair eventually, because the house was built with ramps. And the doorways and hallways were made extra wide to accommodate one."

Did the Kramers fear their ghost? Carl certainly didn't. Geri was alarmed only by the thought that the thing would cross the road and enter her own house. I explained that hauntings are confined to a particular place; that they don't move around. "Unless Buck's the producer of this thing," I said, "then it may go along with him." To which Geri let fly a quick, "He better not be! He's moving into the little farmhouse where I have my office." "And what about my *new* employees?" she demanded. "They'll be here any day now. Good trainers are hard to find; I'm afraid they'll be scared off, too."

"There's nothing to be concerned about. It can't hurt you," I said. "The only harm that can come to somebody is if they become emotionally unstable and jump out the window, fall down the steps, or something like that."

It was patently clear that, even if I'd wanted to, Buck and Mrs. Kramer were much too nervous to go through any type of formal expulsion. So when I suggested they "just let it (the haunting) flow and see what happens," the relief on their faces came as no surprise. My parting advice: "Don't say a word to the new workers; chances are nothing will happen unless they're predisposed to think it will ... that is if somebody tells them about it."

The time came for them to move in and everyone was scrupulously zipper-mouthed. All the same, it wasn't long before they too were bombarded by an inconsiderate house mate.

Geri brought me up to date. In addition to the same old ones, a new noise was breaking the nighttime silence: the sound of a wheelchair rumbling through the hallway and echoing in the cathedral-like great room. And more: "We hired a part time worker.....Tammy Mills, a girl in her late teens. She knew the Niles' boy when he lived here. We were talking one day after things started up again. She told me she used to come to parties at the house. For some reason, she said, she was scared to death of him."

Poor Frank Niles, wasting away in his wheelchair must have been a fearful sight to a young girl. His appearance and possibly his actions contributed to Tammy's discomfort. Geri continued: "She said every time she saw Frank, and it was quite often in those days, he was wearing a white T-shirt. Even in the cold weather he wore one ... something to do with his illness...his high upper body temperature. Then towards the end he was given the choice of having his legs amputated because of all the pain and because they were just dead weight. The last time Mavis saw Frank was just before the operation."

That's what psychic images are like. No messages. No sign of intelligence. I am reasonably satisfied that a traumatically inspired impression of Frank Niles wheels itself around the Kramer ranch--not Frank Niles come back from the dead. If I'm wrong, I'm sure Geri Kramer will be the first to let me know.

Residual Hauntings: Patterns of the Past

"A Phantom In White" points out that all the blame for haunted houses can't be put on discamate (out-of-the-flesh) entities. A good many represent the activities of a force linked to the house itself: a type of replayed psychic memory. In a turn of the century article appearing in Proceedings S.P.R (the official publication of the Society for Psychical Research), pioneer ghost hunter Edmund Gurney hinted at the importance of these "bricks and mortar memories." When commenting on the recurring figure of an old woman seen on her murder bed, he remarks:

> *"The appearance ... suggests not so much any continuing local interest on the part of the deceased person, as (it does) the survival of a mere image, impressed, we cannot guess how, on we cannot guess what, by that person's physical organism, and perceptible at times to those endowed with some cognate form of sensitiveness."*

Modern descriptions of psychic images, or "imprints," likens them to video tape play backs: visual and/or audible reproduction of past events triggered to replay their scenes and sounds. Comparing them to mechanical reproductions is useful when trying to describe the enigma. Just as photographs and tape recordings eventually lose their definition and fade away, imprints tend to diminish after long periods of time: as long as a hundred years or more in some cases. This fact has led to the belief that the only way to get rid of them is to destroy the location in which they manifest. Yet even such as drastic measure as demolishing the house often has little effect on the imprint other than removing the living from their on-again, off-again exposure to it. Psychic imprints tend to stay put. It's rare, but not unheard of, for them to roam the vacant lot where the old house once stood, or inhabit the new one built on its foundation.

And don't think that because you live in a modern house, condominium, or apartment you're automatically exempt from these experiences. Although they're thought to lack the longevity necessary to have acquired impressions, many modern dwellings prove to be just as haunted as the old castles, mansions, and gothic cathedrals of Europe.

Imprints appear on a unique schedule all their own; yet, even when they're idle they permeate the place. They cling tenaciously to the walls, the stairwells, and even the furnishings. No doubt, these memories-in-action are responsible for the apparitions seen generation after generation, always committing the same acts and always seemingly oblivious of their witnesses.

One definition of psychic impressions maintains they are a quirk of nature; the result of strong psychic energy created spontaneously at a moment of great mental and emotional anguish. But I've often wondered why, with all the horrendous crimes committed in the Nazi death camps, more reports of imprinted memories haven't filtered out of Europe since the war's end. Considering the momentous human drama involved you'd think the hills and valleys of Poland, Russia, and Germany would be teeming with the ghastly impressions of those unfortunate human beings. And yet, to the best of my knowledge, there are no recorded incidents of these or any other forms of psychic phenomena in unusual numbers there.

Apparently, not all human tragedies are registered on the atmosphere. If they were, we'd be swamped with them. What is thought provoking is that many of our cases involve replays of ordinary, everyday human experiences and not earth shattering events. Houses in which no tragedy has occurred may acquire an eerie omnipresence in time. Over the past twenty-five years, I've been in hundreds of reputedly haunted houses. Many are warm and friendly, some are not. In houses where murders, suicides, or other tragedies have occurred, a feeling of oppressive sadness sometimes overwhelms you the moment you step through the door. Couple this pervading influence with a healthy imagination and a little knowledge of the houses' history, and you may have an answer for some of the uncanny things reported there.

Can We Explain the Imprint?

Many knowledgeable researchers have pondered the source of these replayed memories. My own theory is that the electromagnetic field encompassing all matter can pick up and record events. In other words, the chemical and electrical activity within the body leaves a "trace" of us in the atmosphere. The surroundings capture this trace and retain it long after we're gone: not dead, necessarily, just gone from the scene. Then anyone sensitive to them would pick them up: see or hear them in their mind's eye.

Thermography (the process of taking heat related infrared photographs) provides us with a convenient analogy. Shots of an empty parking lot have shown cars on the developed film that were there earlier in the day. Some parapsychologists connect these memories with a form of extrasensory perception called Psychometry. According to the theory, all objects are impregnated with psychic energy. Just by holding them, sensitives can "read" them and give accurate details about their past. Theoretically, using the same process, they could "psychometrize" a whole house that way.

In *Operation Trojan Horse*, John Keel, author of a number of works on UFOs and other mysteries, talked about the role primal forms might play in the enigma of extraterrestrial entities:

> *"Throughout history occultists have called these mysterious visitors elementals. There are several kinds in psychic lore. One type is supposedly conjured up by secret magical rites and can assume any form ranging from that of a beautiful woman to hideous monsters. Once a witch or warlock has whipped up such a critter, it will mindlessly repeat the same actions century after century in the same place until another occultist comes along and performs the rite necessary to dissolve it".*

Professor William Roll (State University of West Georgia) speculates that such apparitions are the products of an interaction between the present and past occupants of a house. According to Dr. Roll, deceased individuals or traces of them permeate the physical places that surrounded them in life. These traces may later be activated by a psychically gifted individual. This activation would not necessarily duplicate the past but would be affected by the needs and expectations of the persons seeing the apparitions.

It's an interesting theory and one which may help to explain those rare instances in which imprints seem to respond to the living. There is a premise popular among many researchers that imprints do not directly influence our sense organs: that the nerves controlling optical, auditory, olfactory and tactile stimuli play no part at all in receiving them. We see, hear, smell, feel, and otherwise sense these things, not through the normal organs, but through Telepathy, Clairvoyance, Clairaudience and Clairsentience. Some believe they're closely related to a type of ESP called Retrocognition: clearly seeing events of the past–the exact opposite of Precognition, visions of the future.

Infestation: Mind Constructs on the Loose

Fading memories–death's only certain legacy–cling to the walls and rafters of all houses. Yet there's more to this story than mental impressions and trace recordings. Having acquired sufficient substance ghosts sometimes reach out to punch a hole in the environment of the living.

Parapsychologists define as haunted any place in which varying degrees of psychic phenomena continually occur. My clients use the catchword "ghost" to cover a multitude of scary things ranging from spirit-less, low level physical effects–what I call "infestation," to full blown apparitions who appear in ordinary or phantasmagoric bodies.

In twenty-five years of listening to clients describe their ghostly experiences, I've heard stories about loud knocking and crashing sounds; flying dishes and furniture; clocks that gong without chimes; doors and windows that bang open and shut on their own; doorbells that ring when nobody's there; unexplained flashes of light; appliances and fixtures that turn on by themselves; and yes, even exploding commodes. In fact, one of the most common and irritating complaints is the miscellaneous disappearance and reappearance of "stuff" of all kinds.

Early in the twentieth century, Austrian physicist Ernst Mach developed an abstract mathematical concept to explain the disappearance and reappearance of objects. Mach postulated the existence of a higher dimensionality of space: a fourth dimension where dematerialized objects were kept for varying lengths of time; then re-materialized and returned to their owners.

Modern psychology has a partial answer to these miscellaneous disappearances–one that's a lot less exotic than Ernst Mach's fourth dimension. The annoying habit of overlooking things, even when they're in plain view–right where we left them–can be explained as mental obstructions called "inhibition." It's a type of negative hallucination that causes us to block out those things in our visual field not being concentrated on, and selectively tune out sounds we're not consciously listening to.

Parapsychologists call sudden disappearances, other than those caused by inhibition. "Spontaneous Dematerializations." Objects that disappear mysteriously are called "Apports" in parapsychological jargon, while objects which appear out of no where are called "Asports." The late D. Scott Rogo observed "Such phenomena tend to be either too rare, too disputable or simply too mind-boggling for most researchers to study." Personally, I find that there's no way they can be ignored: my clients won't let me. I generally divide such claims into four distinct categories:

(1) Objects disappear and never return again.
(2) They disappear, then show up later
(3) They're returned to other than normal places.
(4) Objects turn up that no one has ever seen before

Oversights are one thing, but sudden dematerializations, objects that just show up, plumbing and electrical appliances that operate by themselves–that's something else! Most of the time they're just a bloody nuisance; occasionally they herald the arrival of a potent haunting force. I spend a good deal of time trying to sort it out for my callers.

Three out of four of those who call or write me tell of the exasperating occurrences I call infestation. They're a mixture of the real and imagined and, strangely enough, seem to multiply when too much attention is paid to them. It's the sort of thing that makes people say, "If I didn't know better I'd think my house was haunted."

But infestation seldom signals a genuine haunting: the appearance of what is believed to be a non-corporeal entity; a surviving personality. What it announces is the arrival of "mind beyond body": a sometimes random, sometimes directed energy force–one that is capable of knocking us on our cans.

POLTERGEISTS: GHOSTS WITH AN ATTITUDE

All of us have a natural aversion to anxiety and frustration, so much so that we'll do whatever is necessary to keep their end product–stress–down to an endurable level. We may modify our lifestyle by practicing yoga or meditation; get psychological counseling; learn self-hypnosis or biofeedback; use drugs; and even accelerate sexual activity. Whatever works.

A small segment of the population has its own unique method of handling the problem. It's called the POLTERGEIST. Poltergeist is a German word that means "noisy ghost." Unlike haunting entities, which, we will learn, are confined to a place, Poltergeists are "people-oriented": they're connected to a person or persons. Current theory holds that the Poltergeist is not a ghost at all, but the involuntary subconscious creation of a living person, often an adolescent in the midst of puberty.

The phenomena exuded by these biologically and emotionally charged teenagers, almost exclusively physical, are thought to be projected by stress-related psychokinesis. Parapsychologists define poltergeist effects as "externalizations of inner conflicts that could not be resolved in a normal way." It's as if one of the members of the household has sprung a leak in their emotional balloon and is, as a result, inadvertently the cause of all sorts of paranormal effects.

If the various reports of poltergeistic activity have a common link, it is the fact that they all mimic the actions of a youngster in the throes of a temper tantrum. They nearly always display the mentality of an idiot child. In *The Poltergeist* (Nelson Doubleday, Inc., Garden City, NY, 1972), Professor William G. Roll has said, "What we call Poltergeist effects may be an "extrasomatic" expression of psychological stress in the same way that an ulcer is a psychosomatic expression of such stress". Put another way, if fever is the body's attempt to cure itself of a physical ailment then the Poltergeist may be the mind's attempt to throw off an emotional one.

In typical cases, anger and frustration grow within the Agent, the person who causes these projections, and who is also the Focus–the one around whom the phenomena seem to revolve–to the point of explosion. When the "big bang" comes, infestation becomes a full-fledged Poltergeist. Objects appear to be thrown by someone unseen; loud thumpings, deafening hammering and crashing sounds are heard–noises that seem to grow louder and louder by the minute; unaccounted-for pools of water are found; refrigerators, stoves, and washing machines not only malfunction, they, along with other household objects as large as pianos are displaced from their usual locations. A whole host of other effects are witnessed–the worst of which are probably the spontaneous fires–until the event slows and quiets down from a roaring lion to a gentle lamb. It lasts a few days to a few months at most.

Psychokinesis, or PK, is the direct influence of mind on matter. It plays an enormous role in the production of the Poltergeist, as well as being responsible for the physical effects seen in other forms of psychic phenomena. as you might imagine, there are many who argue that PK is scientifically untenable. Yet the evidence for it is so well documented that parapsychologists take its presence for granted. The existence of such a mind-force goes a long way toward explaining the motivating power behind all hauntings.

In a sense, we use mind-over-matter every time we think about, then move a part of our body. PK may be a part of that process, just a bit harder to comprehend. Science understands brain-nerve-muscle functioning, but no one has been able to describe psychokinesis. Like electricity, we don't know what it is; only how it acts and what it does.

Repressed, pent-up hostility finds nonverbal expression through the Poltergeist. The most natural way to vent anger is to scream or physically "punch" it out. The average Poltergeist Agent can't express him or herself in this way. But repressed feelings aren't the only cause of these remarkable outbursts. Recently, researchers have begun to wonder if temporal-lobe epilepsy, nervous system dysfunction, and arteriosclerosis of the brain might play a role in the Poltergeist and in the mystery of hauntings in general. Since (by the criteria established for them by modern parapsychology) Poltergeists are usually the product of adolescence, researchers believe these medical problems may explain instances in which PK exists in places where there are no youngsters.

There's a popular notion that Poltergeists hurt people; an idea reinforced by motion pictures and television. In *Hollywood Ghost Stories*, a television documentary, UCLA parapsychologist Thelma Moss, Ph.D. said, "Nothing in that film *[Poltergeist]* related to what a Poltergeist actually does." The stacked chairs' episode (in which the kitchen chairs were precariously balanced on a table in a matter of seconds) was close; the team of scientific investigators believable, but the rest of the story was pure Hollywood.

These mischief-makers cause untoward fear and uneasiness, but, as far as I am aware no one has ever been seriously hurt by them. Students of parapsychology are quick to point to the so-called "Bell Witch" case as an exception; however, calling this particularly unpleasant wraith a Poltergeist is stretching the definition about as far as the author of *Poltergeist* the movie did.

I admit having seen some minor cuts, bruises, and abrasions caused by flying objects, I've come across dents and holes in walls; doors torn off their hinges; broken dishes, knicknacks, and even furniture demolished, allegedly, by a mischievous ghost. Yet I've never seen or heard of a serious injury having been inflicted by one of them.

I hear numerous cases in which the victims of these pranksters *could* have been harmed but were not. There is absolutely nothing concrete on which to base such a belief, but somewhere there must be a "psychic regulator": an overseer who allows phenomena to be reckless up to, but not beyond the point of really hurting someone. Be that as it may, there is a potentially dangerous side to the Poltergeist.

One case which was particularly destructive (to property) involved a family in the Orland Hills community southwest of Chicago. The story appeared in the *Chicago Tribune* (byline, David Elsner). The apparent Focus of the disturbance was a fourteen-year-old girl who lived with her parents, Bob and Karen Gallo. After spending nearly eight relatively uneventful years in their home, spontaneous fires broke out and a strong smell of sulfur seemed to emanate from the walls. The family was advised to vacate immediately, which they did on April 12, 1988.

Six months later the house was bull-dozed. The rubble was carted off, and the hole that was once their basement filled in–leaving scant trace of what was once the Gallo home. It's a familiar theme, similar to what happened to the Freeling family's home at the conclusion of *Poltergeist* the movie. After twenty-six separate incendiary incidents, eleven of which were witnessed by Orland Hills police and fire department officials, the Gallos were told to tear it down. Spontaneous fires, probably caused by a gas leak, was the reason given by the Travelers Insurance company, which approved and paid for the demolition. Fortunately for the Gallos, Travelers covered the full value of the house. Still, a gas leak was hardly the answer. For seven months, commencing in the spring of 1988, fire department and other investigators; engineers, chemists, geologists, and explosives experts, tried to find a cause. Each ruled out a normal explanation. They considered arson; natural, methane, and sewer gas leaks; faulty wiring, and not-so-practical jokers–then eliminated them all. When summing up his findings, fire inspector Terry Hyland said, "There is no logical explanation. Thank God this [the razing of the house] is the end of it".

And so it goes. We may think of Poltergeists as being timeless, yet, they have an uncanny sense of timing. They somehow know precisely when to start the melee: just as you doze off at night; in the morning, when your eyes are sticky with sleep; or just as you enter the room–off they go. Something, a kitchen towel hanging on the rack; the curtains in the living room; the bed clothing, etc., spontaneously bursts into flames. If you hadn't been present to contain the fire, if it had happened when you were asleep or away, catastrophe might have followed.

Some believe the Poltergeist Focus is a physical medium: a bridge between the world of the living and the realm of the dead. Others speculate that because of their mischievous nature, they may, like imprints, be "elementals": nature spirits or thought forms that draw some of their energy from living human beings and some from the place in which they operate. A few are convinced that Poltergeists are the work of the Devil. I don't attribute a supernatural origin to them. I think there is substantial evidence to support a paranormal source (the subconscious mind of a living person) for all the things they do. If I'm right, then the identity of the "psychic regulator" would be the mind of the Agent. Only under the most extraordinary circumstances would the Agent be unconcerned for the safety of his or her family, not to mention their own well-being.

Yet there's more to this subject than meets the eye. If by Poltergeist we mean spirit-less manifestations of PK (its current, somewhat narrow definition) and not a noisy, spirited ghost (the archaic one given to it 400 years ago), then they're simply random energies run amok. When we analyze them case by case we get a different picture.

In some instances the offending intruder not only violates the Poltergeist nonviolence rule, "Do No Harm," it parades itself around like a genuine haunting entity. For years investigators have identified the Poltergeist as chief culprit in predominately physical disturbances. Yet time and again something puzzling occurs when we study them. Right in the middle of a stereotype Poltergeist scenario replete with psychokinetic effects, up pops a ghost and all the haunting-like accouterments: voices whispering; moans and groans; messages from the beyond, and all sorts of inexplicable sights, sounds, and eerie feelings... On the other hand, so many apparitions and ghosts seem capable of poltergeistic activity (they don't just stand around looking gruesome) that a small but growing number of researchers are beginning to think they're really Poltergeists in disguise.

But apparitions and ghosts are not Poltergeists, not by any modern definition of the term. (In parapsychology jargon an apparition is the image: physical likeness, voice, actions, etc., of a *known* person; while a ghost is likely to remain unidentified throughout its visit.) To consider them as such is to return to the philosophies of the sixteenth century. Nor am I comfortable with the idea of a "smart" Poltergeist. Unless we're wrong about its limitations, I wouldn't grant this mischievous prankster the ability to disguise its true character: a sign of intelligence far surpassing any shown by it in its fifteen hundred year history.

Psychokinetic effects *are* the most often reported; and researchers tend to classify as poltergeist cases any disturbance featuring PK. Yet there are times when our clients are exposed to haunting phenomena even while their noisy intruder was still banging around the house like a kid with a pot and a spoon.

Although establishing exact boundaries is difficult, a line separating strictly poltergeist cases from the haunting-type effects occasionally present in them is not only possible, it's necessary. Defining them separately may not help us solve a particular case one iota, but it definitely helps to understand and keep track of them.

As I see it, the problem started when parapsychologists began lumping two basically different sets of processes into one convenient psychic basket they called "Recurrent Spontaneous Psychokinesis," (RSPK). It's the modern term for Poltergeists originated by Professor William G. Roll. Unfortunately, although Professor Roll disagreed with the idea, for some RSPK became a label for the externalization of repressed teenage hostility only.

In the field, however, we see *two* outwardly similar yet separate forms of phenomena. One, without any sign of intelligence, displays haphazard physical effects: what parapsychologists call "Random RSPK." What distinguishes it from the Poltergeist is the fact that a youngster is not *always* at its center. The other, in addition to recurrent spontaneous psychokinesis, reveals clearly observable haunting phenomena, which are nearly always *directed* toward some specific purpose. The Agent, the one who is the source of these episodes, can be an adolescent or an adult. What separates what I call *"Directed* RSPK "from genuine hauntings is the absence of a disparate human entity or an inhuman *thing* (more on this theme later).

Directed RSPK may show plenty of smarts, but there's no sign that their connected with the dead: they're not supernaturally inspired. On the contrary, there's every reason to believe that they, like the Poltergeist, are created in the subconscious mind of a living human being.

To the orthodox scientist, the existence of PK and its offspring RSPK are radical ideas. One of those who supported the idea early on is Dr. Freda Morris, a California psychologist: "Psychotherapists did not wait for physiologists to develop a complete understanding of the body before beginning to treat psychosomatic disorders. We cannot await physicists' articulation of a cogent theoretical formulation of psychokinesis before we begin to treat families suffering from RSPK." (Source: The July 1974 issue of FATE.)

When we analyze Random and Directed RSPK–divisions of the Poltergeist–we find that:

- Both the non-haunting and haunting varieties are projected by a living Agent who,

- May be an adolescent or an adult (some parapsychologists tend to fix puberty as the arbitrary age limitation for the Poltergeist Agent), and

- In either type of RSPK, Random (non-haunting) or Directed (haunting), adult agents tend to produce less dramatic physical effects than do adolescents. Phenomena produced by teenagers are consistent with what we're used to seeing in traditional Poltergeist outbursts: strong, fitful, out-of-control eruptions of mind-energy. And lastly,

- In Directed RSPK there is an intelligent presence behind the ghostly events missing in Random cases.

History tells us that the word Poltergeist (by classical definition, a noisy ghost or spirit) originated with Martin Luther in the early sixteenth century, and that Hypolyte Leon Denizard Rivail, better known as Allan Kardec, reintroduced it during the 1800s. Today, researchers argue that the term is misapplied since the phenomena it represents are spirit-less projections of repressed pubescent hostility and not the manifestations of *geists* in any form.

Chances are neither Luther, the founder of Protestantism, nor Kardec, the founder of Spiritism intended its use be restricted to the narrow set of rules modern researchers have confined it to. Nor did they attribute its source solely to teenagers. I suspect they chose this compound German word because it best described the phenomena: loud, mischievous behavior by what appeared to be–by its repeated inclusion of entities and haunting-type effects–*SPIRIT* pranksters. On top of that, these complaints often came from homes absent of youngsters altogether. If the circumstances had been otherwise, a term like "Polter*kind*" (German for noisy children) might have been chosen to describe them.

No one began linking them with a living human Agent until the turn of the twentieth-century. By that time the word Poltergeist had been in use for over 400 years to describe a series of mind-boggling distractions that had victimized humanity for the previous fourteen hundred. As a result, even though the term is technically wrong, its modern interpretation is so ingrained in the minds of the public–and especially in the minds of psychical researchers–that no attempt should ever be made to change it. Nor have I done so, except within the context of my own work.

I haven't originated a new concept, just a new term for an old one. In addition to Poltergeists in disguise, parapsychologists have called atypical phenomena: proto-Poltergeists; hidden or camouflaged Poltergeists; or referred to their effects as Poltergeist hauntings.

Parapsychologists have sound reasons for suspecting a youngster in the majority of Random RSPK attacks. Most important is stress: puberty is a time of tremendous physical and emotional change. The demands of blossoming adulthood: sexual maturation and repressed anger and hostility toward parents, peers and society create havoc with the young mind, Yet it's all part of growing up. If Poltergeist demonstrations were the norm, I would be counseling victims around the clock. In my experience, the majority of these kids are shy and standoffish. Some forms of extrasensory perception are more pronounced in the outgoing personality. The youthful RSPK Agent seems to be an exception. Typically, they're around thirteen or fourteen years of age. (They can range from as young as eleven up to the early twenties for "late-bloomers.") Three out of four are young women. Many suffer from an inability to express anger: they bottle it up. RSPK permits them to vent pent-up emotional frustration in a bizarre, yet psychologically healthy way.

Frustration does not end with childhood; we're just prepared to deal with it a little better. There's an immature child that lives within us all, but its actions are fairly well modified by the individuals level of maturity. The emotional and intellectual maturity of the Agent does appear to affect directly the level of intensity and intelligence displayed by the ghost. The Random *adult* RSPK Agent is an older, less boisterous version of the youthful ruckus-raiser. The target of his or her pent-up anger: a spouse, the boss, life's disappointments, ill-health–the possibilities are endless. Otherwise, they're two of a kind, these perpetrators of PK except, of course, one is no longer burdened by pimples.

If adolescent RSPK is the "Big Bang" of the psychic world, then its adult equivalent is the "little pop." Without the pubescent artilleryman there are few mindless torrents unleashed. Directed adult RSPK exhibits some of the dynamics of the Poltergeist, while featuring the phenomenology of a non-supernatural haunting: one without the presence of the dead. For reasons already stated, ample evidence exists that these part-haunting, part-Poltergeists come from a living agent, *not* a surviving personality. Even though they mimic the dead–and in some instances psychic imprints–with some effort we're able to identify them as Directed RSPK. How? For one thing, subliminal messages received during a Psi Session (séance) make little or no sense when examined for signs of the other world. A lack of meaning, though not conclusive, points to a paranormal rather than a supernatural origin. Where imprints might otherwise be suspected there's no connection with a person or prior event (identity of former tenant, their mode of dress, etc.). Furthermore, no reports of ghostly goings-on by previous occupants are on record. (Much more along these lines later.)

In the past, the period of puberty was regarded as "prime time" for the Poltergeist. Nowadays, parapsychologists place less emphasis on unfolding sexuality and its role in adolescent RSPK. Sexual *frustration*, however, may be the cause of some adult cases. Perhaps sexual milestones like the "Seven Year Itch" and the "Mid-life Crisis" are responsible for touching off adult RSPK in the same way it was thought puberty fired-off the Poltergeist. Naturally, there are crossovers in this scheme. At the end of the spectrum are the fairly tame projections of precocious youth; at the other, those of the elderly going through a second childhood. This time around the caliber of their cannon may be a little less impressive, but it's still plenty noisy. In recent years, with ever more senior citizens obliged to move in with their children(and with financial problems forcing their adult grandchildren to live there too), we've observed and increase in RSPK cases: the emotional strain of the generation gap the probable cause.

In my opinion, both Random and directed RSPK are cries for help generated by psychological mechanisms. Both represent novel, yet effective "safety valves" for venting repressed hostility and frustration. Once we recognized the part played by grown-ups in the process, the most puzzling aspect of it seemed to fall into place: the fact that Poltergeists could be observed in homes where there were no pubescent youngsters.

Extensions of the Living Psyche

I classify the things I study as: Imprints, Poltergeists, haunting ghosts and apparitions of the dead, and forces that mimic them but prove to be projections of the living. All, except perhaps the imprint, have one thing in common: regardless of their genesis–whether thrown at us by a living mind or a supernatural being–I have come to believe that haunting effects are created by what researcher F.W.H. Myers called "persistent personal energy," and are projected through the *mind of a living human being.* It is only their inceptive source–their origin that may be dissimilar. That source is either:

1) The subconscious mind of a living person (the region below conscious awareness). Or,

2) An outside, non-subjective energy using the mind of a living person as a *conduit* through which phenomena are externalized.

Because the concept of "mind beyond body" is so crucial to an understanding of what I do when I do my thing, I will reiterate: *It is the living subconscious mind that projects all haunting phenomena into the environment, regardless of who or what the mysterious force is that initiates it.* In response to some need (example number one), the mind reaches out to the physical environment on behalf of itself. The phenomena that ensue may be scary as hell, but they aren't supernatural–not in the sense of an unnatural act. They're not beyond Nature, only our current understanding of Nature.

In the second example, an objective energy source: a ghost, apparition, or "entity," makes use of a living person's mind to manifest itself. It is this concept, smacking as it does of the supernatural, that sometimes drives parapsychologists to opposite poles of opinion. But regardless of the underlying cause, most researchers agree that without the presence of a living human catalyst it would be impossible for "ghosts" to perform.

Of course, all this is theory. But subconscious externalization is not a new idea. Those few scientists who are willing to concede that some haunting incidents really happen have been saying for years that they're the offspring of mind-projections: internally generated subjective creations, and PK powered energies. (Remember that PK is the ability to create a disturbance at a distance without the use of the physical body.) In *Paradoxes of Everyday Life*, Milton Saperstein tells us that our unconscious "is like a vast subterranean factory that is never idle, where work goes on day and night from the time we are born until the moment of our death."

Mental life is managed on the conscious and the subconscious planes. Everything that occurs in our life, awake or asleep, is stored away in the subconscious mind–our subterranean factory open round-the-clock. In addition to being a "launching pad" for PK, it's a huge storage area for every one of our experiences, thought, feelings and impressions–not all of which are consciously remembered.

Robert H. Ashby, author of *The Guide Book for the Study of Psychical Research* (Samuel Weiser, New York), defined ghostly forms as hallucinations with some paranormal cause. He wrote:

> *"Although some students hold that some apparitions have a spatial reality...most researchers feel that an apparition is an exteriorized hallucination corresponding to some veridical person or occurrence."*

I believe they're the result of intense internal imagery; that they're designed and constructed from mind storage, then sent out in three dimensional form as "genuine sensory hallucinations." They may be the likeness of one or a combination of different people or things put together from material previously stored away. They may be seen, heard, touched, smelled, and probably even tasted. Their presence may raise the hackles on the back of your neck, or go completely unnoticed.

Descriptions of some apparitions, their actions and even the clothing they wear, gives added weight to the mind-construct theory. I often hear that they appear"...just the way they did the last time I saw them"; or, ".....exactly how I remember them when I think back to those times." Or I'm told they look the way they did in a dream. Yet these experiences aren't totally subjective. Often they're supported by others who have the identical, or nearly identical hallucination.

At the same time, not many confuse them with physical beings, even if they appear to be substantial. Witnesses perceive them by means of a "shared collective hallucination" produced through auto-suggestion, or by their own ability to receive and the hallucinator's ability to send telepathic picture forms. Telepathy in pictures is similar to the transmissions of a miniature television camera within the brain of the sender to its counterpart, a tiny TV set in the brain of the receiver. In Arthur Koestler's *The Roots of Coincidence* (Random House, New York) , the author tells us how easily images could be sent symbolically from one person to another by means of particles called neutrinos.

Just as mind-over-matter is offered to explain the Poltergeist, mind-constructions may account for the appearance of ghosts. For most, extensions of the living psyche explain their seeming reality quite nicely. To those who've witnessed them in three-dimensional living color and stereophonic sound it's tantamount to a punch in the eye. When the suggestion was made to one of my clients, he fired back, "Hallucination my *ass*! We didn't *imagine* this stuff, it was real!"

Some haunting entities do seem to have an objective reality. In addition to the fact that more than one person at a time has witnessed them, their images and sounds have been recorded; they're described as solid and very much alive looking; and they can affect and are subject to the environment they visit: moving things around, avoiding objects and people in their path, and even showing up as reflections in mirrors.

Notwithstanding my client's rear-end, hallucination (sensing things that aren't there) play a major role in hauntings. Today, the word suggests some sort of drug-related "trip" or psychological disorder. For our purposes it is defined as "Any supposed sensory perception that has no objective counterpart within the field of vision, hearing, etc." It's like dreaming while your wide awake.

A genuine sensory hallucination is the actual sensation of seeing or hearing persons and objects that are not really there. The mechanics of the experience differ in no way from the authentic thing, except that the sense organs have nothing to do with their production. The mind creates sensory hallucinations under the influence of highly charged emotion, similar to that present during most hauntings.

According to a survey by the Institute of Psychophysical Research in Oxford, England, those who tell of hallucinatory experiences are "otherwise perfectly normal and have no more psychiatric history than the public at large." (Source: Morton Schatzman, *New York Times.*)

Chapter 2: The Haunting Entity

"For over all there hung a cloud of fear, a sense of mystery the spirit daunted, and said, as plain as whisper in the ear, the place is haunted."
(Thomas Hood)

We human beings are, by our own reckoning, at the top of the evolutionary chain. We're the best nature has to offer. Even so, lower forms of life have one distinct advantage over us. As far as we can tell they are blissfully unaware of their mortality, while we must knowingly bare the "Curse of Eden." During all except perhaps the last of life's stages, indulging in the death denying game, *It's not going to happen to me,* is psychologically healthy. Everyone does it. For many, the existence of the ghost--like a strong dose of religion--offers the promise of avoiding oblivion.

Such a hope is the haunting entity, defined as the surviving personality, essence, astral form, spirit or soul of someone who once lived. In its present condition it can be seen, heard, or impress our senses in other ways. It's thought to be an objective source of energy with an intelligent purpose behind its actions. It is the rarest of rarities in psychical research.

Important in determining if a disturbance is something more than an impression left behind, or psychic energy gone awry, is the evidence of individuality, intelligence and intent displayed. Genuine haunting entities routinely exhibit these characteristics.

But what do we call those who awake from the last sleep? Parapsychologists and amateurs struggle with a name for the dead, or "undead" if you prefer. Some call them discarnate entities, returnees from extinction, or out-of-the-flesh beings. Researcher Loyd Auerbach likes the term "dead guys," and uses it frequently: that's descriptive enough. I've thought of calling them 'Supernaturals" then "Paranormals" would be entities created solely by the living mind; and "Naturals," naturally, would be misinterpreted things that have nothing to do with parapsychology.

It's confusing, but all creatures, large, small, and ethereal deserve a name. And, in this narrative, I've used "Genuine Haunting Entities" to describe these alleged returnees. It's pompous of me, I know, to call them "Genuine"; yet, to the terror-stricken people who report them--the name more than fits.

The consensus is that Genuine Haunting Entities are "place-oriented": they haunt a certain location and remain there for long periods. Though not typical, there have been cases in which something resembling attachment was formed (the entity for the individual, or for the whole family), and it followed them from house to house--even from city to city. Inhuman entities: *things* that have never lived in human form, are said to obsess, or harass a particular person or family wherever they go. Thankfully, there's little evidence on these gruesome characters to support their existence, and much to indicate a psychological source for them. (It's a comforting, if not convincing argument when we're in an evil plagued house on a lightning-filled night.)

Included in a few of my case histories is circumstantial evidence, but no proof, that the human varieties represent the dead. If man does survive bodily death and returns to haunt his earthly environs, he would need a source of energy to make his presence known. Most researchers believe that any theory of survival is dependent upon the ability of that source to create its own effects without the help of a living catalyst. There's not much evidence of it in the field.

Some authors write about hauntings as an acceptable, even verifiable basis for immortality. This is not such a book. It's not about life after death, though it may shed a bit of additional light on that possibility. I have no universal truths or profound insights to offer the reader on survival. For myself, I have more questions about the nature of immortality now than I did more than twenty-five years ago (when, I admit, the subject was far less a personal issue). It might make my work more exciting, but it would be intellectually dishonest to claim otherwise. Since the 1880s and the founding of the Societies For Psychical Research, a special group of investigators has emerged to test that "middle ground between science and superstition." The Ghost has been studied in all its aspects by scientists esteemed as mental giants in their fields. Few encountered more than a trace of their quarry in person. Psychiatrists, psychologists and philosophers; physicists, astronomers and mathematicians; physiologists, physicians, chemists, and dedicated amateurs have taken up the challenge. Their lives were devoted to pondering the complexities of the subject. Still, there are no concrete answers and no single theory that can account for their existence.

Nor has Psychical Research come upon anything approaching irrefutable proof that survival takes place. All they have to show for a hundred-plus years of intense investigation is an accumulation of evidence suggesting that life continues beyond death in a meaningful way. The material collected worldwide is voluminous, yet it has failed to convince the nonbelievers, leaving immortality a matter of personal belief.

Whether spirits truly exist or survival is a fact, the effects produced during genuine hauntings are real. Of that I have no doubt. I hope to show that the energy used to create these effects resides within the subconscious mind of a living human being. Through subconscious externalizations this person, usually a member of the haunted household, unwittingly provides the power necessary for all the extraordinary things that occur.

Case File: Deathbed Promises

"What the mother sings to the cradle goes all the way down to the coffin."
(Henry Ward Beecher)

If Genuine Haunting Entities--the human variety--are indeed death's survivors, why do they come back? More particularly, why do they come back so full of noise? The best guess is they have an overwhelming need to be heard. There's a message for someone and they're bursting their ethereal seams to deliver it. The more commotion they make the better the chance they'll get through. Message bringers, no matter their source, nearly always have something important to say--leastwise as far as they're concerned. Presumably, they intend to say it in person; sometimes they can't, and that's where I come in. I open a channel of communication for them.

In the fall of 1992, what began as a succession of typical Infestation pranks evolved into one of the most significant adventures I have on record. To family and friends there was little doubt that the love of a mother for her troubled son had extended beyond the grave. It was an explanation appealing in its simplicity. The investigation was prompted as so many are by a series of frantic telephone calls. Since I have guaranteed anonymity to all those involved I'll refer to my caller as Joan Snyder.

The first of Joan's pleas for help came on the third of October. At the time she was thirty-eight, a homemaker, mother, and grandmother, living just across Tampa Bay in St. Petersburg. She made a point of telling me that her husband Larry (thirty-five) "isn't keen on asking for your kind of help." It seemed that Mr. Snyder felt any outside interference would brand him as some kind of weirdo. Without consulting him, she'd asked for it anyway:

"The thing is," she began, "for a couple of weeks now we've been hearing these banging sounds coming from the garage ... like someone's beating the concrete floor with a shovel. It wakes us up around two, but stops the minute you open the door to go in there. When you get into the garage, it starts up again ... only now it's coming from inside the house. "And the flies ... thousands of flies. We sprayed but it didn't help. I saw this movie about flies in a house. I know it sounds crazy, but it's the main reason I called you."

Larry and Joan Snyder were renting a three-bedroom ranch across the street from a church cemetery. Joan's eighteen-year-old daughter Tricia; Barry, Tricia's two-year-old; Heather, sixteen, and her friend Tony, seventeen, lived with them. Tony slept on a couch in the living room. The girls were Larry's stepdaughters: Joan having been married previously.

Sometimes a series of phone calls leads to an on the scene investigation. More often, callers learn to handle their problems themselves by waiting for them to subside. Events at the Snyder's progressed so ominously that they were unwilling or, more accurately, unable to wait. By the middle of October they were experiencing a rapidly expanding haunting. On the sixteenth, Tony, Heather's live-in friend, was jolted awake by the sound of a woman moaning:

"It's happened twice now, and it really shakes him up," Joan observed. "Once, his clothes were tied in a knot. Another time, the T-shirt he was sleeping in had holes burned in it. Everyone in the house smokes except Tony ... and, of course, the baby. We don't hear banging anymore. Now it's furniture moving around in the garage. I know you told me to ignore all this. I've tried ...I really have. But last night Heather, my unmarried girl, was taking a shower and she heard somebody humming up a storm right along with her. And Bucky, our mixed spaniel, he won't go near the garage. And all my husband can contribute is, 'Hey, I don't want to hear about it!" "Is he exposed to any of this?" I wanted to know. "Sure. He hears things just like the rest of us. He just says he's not superstitious and that's the end of the conversation." Again, I assured Joan that there was nothing to be afraid of. If she and her husband were interested I'd come up and have a look. "No. Unh-unh," she said cautiously. Larry's not ready for that... not yet, anyway." "Okay, but if you need to talk to someone you can reach me anytime, day or night."

Eight o'clock the next morning Joan was filling me in on the night before: "I don't know how much of this I can take. People have a limit! Last night we both laid there in the dark listening to the furniture bang around downstairs; neither of us with enough guts to go and see. I said, 'Larry, don't you hear that?' And he goes, 'Uh-huh,' and he just lays there with his eyes closed. "Finally, I got up. Everything in the living room ... all the furniture and everything was piled in a corner; and Margie's picture [her mother-in-law] was turned around facing the wall. The girls came down and were helping me get things back in order, when all of us heard this voice say, 'Good Morning ... I'm just doing this for the hell of it.' "You think you could come out ...I mean if Larry says it's okay?" "Whenever you're ready for me."

I didn't hear from her again until Wednesday, the twenty-seventh. "I should have called you sooner," she said nervously. "I would have but Larry was against it." "Mrs. Snyder, I must tell you that unless both you and your husband agree to use my services there's not much I can do for you ... not directly, that is. I'll continue to try to help you over the telephone, of course." "Wait! Just let me tell you what we're going through. The noises ... they go on almost every night; the furniture we find scattered every which way. Larry's mother's picture comes off the wall and sails across the room." "Has anyone actually seen that?" I asked. "No. But it was clear on the other side." Joan went on: "My daughter Tricia's little boy...my grandson Barry...he keeps pointing and saying 'Minnie, Minnie.' Minnie is what she calls my friend Mimi Garrett. Mimi adores that baby, and Barry loves her back. He keeps saying it even when there's no one there. I went in to him this morning and he was standing up in his crib pointing away from me and calling, 'Minnie!' He reached out his hands and jumped up and down; you know, the way a baby will do when he wants to be picked up. But he wasn't reaching for me, and I was the only other person in the room."

There was more. On Monday, her sister Bonnie arrived to find the front door wide open. Naturally she assumed someone was home and called out, "Sis...where are you?" From the garage she heard a voice answer, *I'm in here. Come in.* She moved to the garage door, reached for the knob--then stopped in her tracks. The latch was still in place. The door was locked on the kitchen side. Since there was no other route in or out of the garage, the only way her sister could be in there was if someone had *locked* her in there. Before Bonnie could react, Joan came bouncing into the kitchen carrying groceries. "Hi. How'd you get in here?" "I walked in the front d....Oh, my gosh! Someone's in your garage."

Nearly frantic, they spent the next twenty minutes searching the house--top to bottom. "There was nothing!" a still shaking voice informed me. "I spoke to Larry," Joan continued. "He's not too thrilled about it but says it's all right for you to come out." I wasn't all that thrilled, either... The idea of trying to work around an uncooperative spouse wasn't appealing; but the situation warranted a house call. "I'll be there Saturday night (October 30)," I said.

Eight hours before I was to arrive, Joan was on the line again. Sometime during the night (she failed to check the clock) they'd heard the "most awful screeching sounds." When Larry investigated, the door to the garage was lying on the kitchen floor. "When you get here tonight," she said, "I want you to take a good *look* at the screw holes in the door frame. They're stripped The door was torn right off the jamb!"

Armed with my tools and devices, I arrived at 7:30 p.m. All the principals were there to greet me except our reluctant host, Larry; who picked that evening to visit his sister. Joan apologized. "It's just as well. He's been so irritable, snapping at everyone, ever since he heard you were coming." Often, the man of the house will detach himself from the haunting, from his own personal fears--and even from the family he loves.

Joan, her daughters Tricia and Heather, little Barry, his aunt Bonnie, and boarder Tony, led us to the garage. Bucky, the family's mixed spaniel stayed glued to the doorway. Nothing in this world could have induced him to join us. Joan, as though reading my thoughts, whispered, "Even when I carry him down he can't wait to get away from me and get back into the house." The garage had been converted to a den and was furnished with a worn sofa, occasional chairs, end tables, coffee table, console stereo, and an old beat up TV--all resting on a threadbare rug. Tony carried a couple of folding chairs down with him. While Barry played quietly on the floor, the rest of us gathered round my tape recorder. For the benefit of the archives, each of the witnesses testified to their role in the haunting. It was, for the most part, a reprise of what Joan had told me over the telephone.

Tony often heard tapping sounds coming from the garage; and "a tire gauge floated across my hide-a-bed one time." Heather complained, "When I get into the shower it feels like someone gets in there with me."

Everyone agreed that objects around the house disappear, then reappear in odd places. Unexplained footsteps on the stairs, clicking sounds, banging and rattling noises, and loud thumps were added to the list. Joan told me that the furnace, humidifier, air conditioner, washer, dryer, and hot water heater were checked out as possible sources: all were in good working order. "And even if they weren't," she protested, "it couldn't account for the furniture ... could it?"

Patting the armrest she said, "One thing I didn't tell you happened right here on this couch. It was on a Sunday. We were sitting here, Tricia, Heather and me, watching television. This end lifted up in the air and came back down hard. (Both girls nodded at the same moment.) You can bet we got out of here as fast as we could."

Tricia confirmed that Barry calls out for "Minnie," then offered a rather perceptive observation. She wondered if there could be a connection between the baby calling to Minnie and the fact that Mimi Garrett (alias Minnie) has red hair.

Margaret Snyder, Larry's mom, had had red hair, too. On the wall behind the sofa, in a damaged frame, hung an oil painting of Larry's mother. "Maggie," as everyone called her, passed away in 1989, the year before Barry was born. Her flaming hair was the outstanding feature of the portrait.

The elder Mrs. Snyder had been a multi-talented woman: "She was an artist and a writer," Joan boasted. "Painted her own picture and wrote a biography of herself." Handing me a ceramic bust, she added, "Maggie made this statue of our Savior; that's her own hair on it." The work, especially the straggly dark red hair glued to the head, instead of having some artistic merit struck me as grotesque. "You know," she said thoughtfully, "I think it's grown. I don't remember it being this long and stringy before."

I'd been there for nearly an hour. There was no question that the haunting had progressed rapidly. The most recent occurrence, the garage door torn off its jamb, was one of the most dramatic exhibitions of attention-getting PK I'd ever come across. Just after eight-thirty we climbed the steps and set up shop in the dining room. Bucky, the not so intrepid pooch, was put outside. After tucking Barry into bed, Joan joined the others around the dining room table.

I placed the newsprint pad and writing planchette-- tools of communication--in front of them, then dimmed the lights. In silence we waited as our eyes adjusted to the semi-darkness. Apprehension was building. I began with a relaxation technique: "Sit back in your chair, uncross your legs, untie or loosen your shoes, and close your eyes. Begin to breathe deeply, and with every breath you exhale you will become more deeply relaxed. Breathe in and out. Breathe in ... and out. In...and out. Now concentrate on the weight of your shoes. Your shoes are foreign to your body ... they begin to feel heavy, and this heavy relaxation from your toes to your heels to your ankles becomes very noticeable. You are now feeling this heavy relaxation moving upward into the calves of your legs, feeling the weight of your legs pushing down, heavier and heavier-and feeling your legs deeply ... deeply relaxing, and this heavy relaxation moves into the knees as you concentrate only on my voice. Pay no attention to any sounds except my voice, for these outside sounds can't distract or disturb you, but will tend to relax you and allow you to go even deeper into this deep, heavy relaxation."

Using a non-musical variation of *Those Dry Bones,* I worked my way up the body: "Now feel the relaxation moving upward into your thighs and hips and through the mid section of your body. Feel the stomach muscles relaxing, deeply relaxing, and the entire chest area becoming saturated with relaxation. Breathing becomes very deep, gentle and rhythmic, and the drowsy, sleepy, daydreaming feeling of relaxation takes over ... drifting you down, deeper and deeper. Your arms, hands, and fingers are relaxing, feeling a numb, pleasant, tingling sensation through your fingers as this relaxation grows deeper and deeper. Deeper and deeper. Neck muscles are relaxing and all the little muscles in the scalp are letting go, feeling as if the blood is circulating very close to the skin. This relaxation moves down over your forehead and down over your eyelids like a dark veil of sleep as your jaw muscles relax deeply ..... deeply relaxing, relaxing, and growing heavier.

"Now that you are completely relaxed," I continued softly, "open your eyes and place the fingers of your non-writing hand on top of the tripod. Use the opposite hand to the one you write with." As soon as they did, I rested my fingertips briefly and ever so lightly on each of the participant's hands to reinforce the command. While moving from hand to hand, I said, "I'm going to ask questions about the disturbances in this house. The writing planchette will supply the answers. I will ask each question once. You: Joan, Tricia, Heather, Brenda, Tony ... you will repeat the question to yourselves. You will remain comfortable and relaxed all through the proceedings." After a quarter-hour of lines, circles, and figure eights, the planchette stood poised for action:

Andrew: "If there's an unseen presence in this house identify yourself now!"

Reply: (None)

Andrew: "You know that unless you identify yourself you must leave this house immediately, never to return."

Reply: (None)

I repeated the instruction. There was no reply.

Andrew: "Are you responsible for the disturbances in this house?"

Reply: "Yes"

Andrew: "What is the purpose of these disturbances."

Reply: "For me. Me."

Andrew: "Is there a message for someone in this house?"

It took five minutes for a reply to come through. The message, written in lower case cursive letters, read "you must move": a rather common demand from haunting entities upset by those who are living in *their* houses. The words "you" and "must" came through quickly and were easily read. The third word, "move," however, was more difficult to decipher. It could have been interpreted as want instead of move. The m. began with a flourish--a loop that made it look like a w. (The m. in "must" did likewise, but was clearly *not* a w.) The o. could have been an a. Just as scripted o's and a's are often similar, v's and n's--if not written clearly--may be confused for one another. And, at the end, the small e. had a line through its middle, making it look as much like a t. as an e.

As if there wasn't enough to befuddle us, the message stopped abruptly after the third word. "You must want" was an incomplete sentence; it didn't make sense.

"You must move" was not only complete, it was, for all we knew, appropriate to the situation.

Andrew: "Is the message, 'You must move?'"

Reply: ('Three large X's were drawn.)

Andrew: "What is the message, then?"

There was no doubt that the next words were "TO" and "LIVE" spelled out in capital letters and repeated twice. The completed message was either, "You must *move* to live," a threat; or they were words of inspiration: "You must *want* to live." "You must *want to* live!" said Joan in a trembling voice. "Sounds like something Maggie would have said. She was a strong-willed woman. If anyone could come back, Maggie could. I'm calling Larry." In a few minutes her disgruntled husband was back--riled, no doubt, by the summons. After seeing what was on the pad he all but accused me of trying to bamboozle them. I hadn't even touched the planchette!

"What's *this* supposed to be?" he asked incredulously. *Why would he react that way,* I asked myself, *unless the message meant something to him?* My instincts told me it did. Larry sat down and took a closer look. Still frowning, he laced his hands behind his head and, with what sounded like gravel in his throat, said, "I don't know about all this. I don't believe in *any* of this." Then, scratching his chin and halfway smiling, he added, "But that sure sounds like my poor mama ... even to the way she used to make her W's and T's." What followed had completely slipped my memory until the tape jogged it back: Larry spent the next half-hour looking for his mothers hand-written autobiography. He wanted to show us the similarities in the writing: "I can't find it," he said dejectedly.

"In any event," I offered, "it wouldn't prove anything one way or the other." "My mother had a stroke. She was only fifty-six when she died." "If it is her, Mr. Snyder, do you think the message is for you?" I asked. "Oh, yeah. It's what she said that night at the hospital."

The words came slowly as he related the story of Maggie's passing. She'd been in the intensive care unit for several days when he and his sister were called to her bedside: "My mother was always after me about my drinking. I'm what you call a problem drinker." He glanced at Joan, stiffened a little, then admitted, "Okay, I'm an alcoholic!" Larry was still recovering from the loss of his left eye (a direct result of alcohol abuse) the night he rushed to Maggie's deathbed. The last earthly concerns of Margaret Snyder had been for the welfare of her son. Before she passed away she made him solemnly promise to give up drinking:

"She used to say to me, 'God gives us heartache and the devil gives us whiskey.' "Even in all her pain she told me I had a lot to live for. She said, 'You must want to live. Grow up and be a man, son-do it for me.' "She whispered it in my ear. A little while later she was gone. That's the first time I've talked about it, except to my sister. You know, it's funny. We keep hearing these noises around two in the morning. My mother died at two in the morning." Her son had kept his promise. He went on "the wagon" and stayed dry for over two years. Then money worries and problems at work overwhelmed him; he was back on the bottle: "I think it was the same week-the last week of September when my doctor said I wouldn't see forty if I didn't stop. He said my liver was about shot. It all started then. "Do you think that'll be the end of it. I mean will these things stop now?" he asked. I answered his question: "It has no reason to go on now that you've got the message ..... unless you don't listen to it." "Yeah, I guess."

At 10:00 p.m. we resumed the seance. This time Larry joined his wife and step-daughters at the table, but not before calling his sister to get her opinion (and, I suspect, her approval as well).

Again, I requested that the communicating source identify itself. This time the name "Margaret" was written, followed by the words, "LIVE ... LIVE." At a quarter to twelve the planchette no longer responded to my questions. The seance was over. I told the family they would have a feeling of well-being and would remember everything that took place that evening.

But what had taken place that night before Halloween? Our clients, even Larry, were satisfied that the spirit of Maggie Snyder had paid them a visit. There was circumstantial evidence to support it: more, I believe, than in any other case I'd been on. And yet, it is equally possible that simple self-preservation caused the outbreak of phenomena. In order to jolt his conscious mind into recognizing and then doing something about this life threatening situation, Larry's subconscious may have resorted to psychokinetic dramatics. The noises and physical outbursts may have been nothing more than

involuntary mind-projections: a last ditch attempt to stave off personal disaster. But if Larry was the source and the catalyst how do we explain those haunting effects that occurred when he was absent from the scene. Even the evidential message was first recorded while he was at his sister's house.

The answer lies, I think, in the fact that in some RSPK cases the Agent is not always present when things "bang" and "pop." Researchers have posited that some type of residual energy “bleed-off” remains in the premises even when he or she is not there.

This case emphasizes, as well as any I've studied, the enigma behind hauntings: Supernaturalism versus Paranormalism. It is the crux of ghost research. In my experience many so-called genuine hauntings, complete with all the trappings of a materialized entity, turn out to be nothing of the sort. They prove – at least to my satisfaction – to be the subconscious creations of a living human being in need of a method of expression, yet unable to find a socially acceptable way of achieving it. On the other hand, even though it is virtually impossible to rule out a subconscious and therefore natural explanation for reports like these, I'm sure the reader will agree--they do include compelling evidence in support of life after death.

Ultimately, there are two possible resolutions to this seemingly supernatural tale. The most obvious, of course, is that Larry's mother returned from the grave to scold and to save him. The most logical, considering what we know about Directed RSPK, is that he found a way to save himself.

Alcoholic dependence brought him to the brink of death; the message, on top of the startling psychic phenomena, brought him to his senses. Whatever force was responsible for triggering the haunting, when Larry Snyder stopped drinking, it stopped, too.

I don't want the reader to get the idea that all seances result in meaningful messages. What is communicated is not always as profound as "Maggies" rebuke to her wayward son. Time and again the intensity of the disturbance is out of proportion to the content of the message. Like the translated words of an Italian opera they suffer and fall short of their soul-stirring score. Messages without significance: lifeless, mundane utterances seem to parallel Imprints that lack profundity. Why bother? Why bother to deliver boring messages? For that matter, why bother to replay correspondingly commonplace events like walking down a staircase, or sitting stone-like at a bar? Who cares? Who needs it?

A reason given by some survival theorists for the absence of clear and specific messages involves the state of one's consciousness in the afterworld. According to them, the shock of death can impair the memory. It's a dream-like existence, they tell us, in which physical intrusions (images, sounds, aromas, etc.) into the world of the living are possible, but lack intelligent direction. The omission of meaningful dialogue between ghost and attentive listener is sighted by others as a sure sign that a living person rather than a dead one is responsible. You need only recall the ramblings of the Ouija board to follow their reasoning.

Further support for a living source comes from the fact that during the seance it is often difficult for a communicating entity to comprehend that he is no longer a part of the living world. Perhaps it's because he has not yet *left* the living world.

Case File: A Grandmother's Warning

> *"I cannot tell how the truth may be; I say the tale as 'twas said to me."*
> (Sir Walter Scott)

Like most Americans I was glued to my television set during the opening days of the War in the Persian Gulf. On January 19, 1991, Cable News Network reported the second Iraqi "Scud" missile attack on the city of Tel Aviv. Israel's largest city was asleep when the alarm sounded. Quickly its citizens scrambled to their "safe" rooms, donned gas masks, waited and prayed As the CNN commentator spoke, the camera scanned the nighttime sky. Suddenly, we caught a glimpse of a fiery inbound rocket. "It's going to miss us," he said, ducking his head a little. And it did, crashing into the sea. But it wasn't a Scud this time. This Russian built vehicle was the kind that delivers cosmonauts not catastrophe. A decaying satellite booster had reentered the atmosphere, arched over the Jewish State, then fizzled harmlessly into the Mediterranean. The irony of the event was lost in the excitement. Of all times and of all places for a spent sputnik to come plummeting back to earth. Why now? And why Tel Aviv-the pre-announced target of Saddam Hussein's vengeance?

Carl Jung called coincidences like these "synchronistic." One day, the Swiss psychologist was treating a patient who complained of recurring nightmares in which she was plagued by a golden scarab beetle. As he listened, he was distracted by a tapping sound coming from the window. He turned round and saw a flying insect knocking against the windowpane from outside. He opened the window and caught in his hand a large scarab beetle as it tried to come in. That's synchronicity!

Mere chance incidents like these fly in the face of logic. Jung believed they represented more than chance; that synchronicity suggested "a kind of harmony at work in the interrelation of both psychic and physical events"; that, "...occurrences similar in form inevitably attract each other." "The concept of synchronicity," he wrote, "indicates a meaningful coincidence of two or more events, where something other than the probability of chance is involved.." (Carl Jung, quoted in *The Journal of Religious Thought.)*

In June 1984, I visited a thirty-three-year-old divorcee and her three children in rural Tennessee. Her teenage son, Tommy, had a lollapaloosa of a *meaningful coincidence.* Telephone conversations with Janet McDonald, a name I'll use to conceal his mom's identity, seemed to point to RSPK as the cause. There were equally strong indications of the presence of a Genuine Haunting Entity in the form of a guardian angel. During my visit, Mrs. McDonald repeated her story for me. In addition to Tommy (fourteen), daughter Jessica (twelve) and Wanda (nine) were also there:

Janet: "Tommy's father ... he's lived in Memphis since our divorce three months ago ... he sent him a treasure finder [a magnetic metal detector] for his birthday. He was out playing with it a couple of weeks ago when he dug up a tin box buried in the shack. There was some snapshots in it of people, like from the thirties, you know? One of them I recognized. It was my mother when she was a child. I'd seen other pictures of her when she was little, so that's how I knew it was her. An older woman had her arm around her, and I just guessed that was my grandma. I never saw a picture of her before ... she passed on when I was still little."

Andrew: "Is your mother still alive?"

Janet: "Yeah. She's in Knoxville with my brother. I sent her the snapshot, but haven't heard back yet."

Janet: "The night after Tommy dug the tin up, he and the girls saw a woman all aglow come into their rooms."

Andrew: "Can you describe what you saw, Tommy?"

Tommy: "It was a lady with a white dress on, and a light coming out of her. They saw her. Jessie (daughter Jessica) cried, and Wanda pulled the covers over her head."

Andrew: "Is that right, girls? Is that what you did?" [Neither attempted to answer.]

Andrew: "What did the lady do?"

Tommy: "Nothin'! She just looked at me and pointed her finger like this [pointing a crooked index finger at me]."

Janet: "I thought they dreamed it. Then I thought they made it up because they all saw her. But, then, the next night it happened again."

Tommy: "She leaned over me and pointed at these beads [holding them up for us to see]."

Janet: "Tommy found this string of metal beads in the shack with his detector."

Tommy: "I hung 'em on the wall next to my bed. The lady was pointing at *them* and lookin' at me."

Andrew: "Did she say anything to you?"

Tommy: "Yeah. She didn't say it, exactly, but I heard it. It was like, Watch out you child, or somethin' like that."

Janet: "He told me she said, 'Beware, child of my child. He couldn't have made *that* up."

Andrew: "What do you think she meant by that?"

Tommy: "To stop going to the shack, I guess."

Janet: "Well, we didn't know that ... not then."

Andrew: "Did you recognize her?"

Tommy: "Yeah. She's in the picture."

Janet: "When I showed him the snapshot of his grandma, he said the lady squeezing her was the glowing lady ... which would be his great-grandma. "

Tommy: "Only she looked better when I seen her in my room."

Janet: "He means she looked younger."
Tommy: "Uh-huh."

Andrew: "Are you from around here, Mrs. McDonald?"

Janet: "Not far. I was born about ninety miles from here. We lived on a farm till I was seven, then moved to a little town you probably never even heard of."

Andrew: "Can you think of any reason a photograph of your mother and grandmother would be in that shack? Did they ever live around here? Was there anyone else in those pictures you recognized?"

Janet: "No. Unh-unh."

Andrew: "How do you account for all of this?"

Janet: "I guess I can't account for it."

Janet: "Right before grandma warned us about the shack– I'll tell you about that in a minute ... we had poltergeists. The kids were playing in the old truck [a rusting 1953 Dodge pickup parked to the left of the driveway] when it started rocking up and down and back and forth so furious they couldn't open the door to get out."

Andrew: "Did you see it?"

Janet: "I heard it ... squeaking and rattling first; then I saw it. After that, for about a week the house was hit by rocks ... hundreds of them on the roof and against the sides. They're piled up out back. When you picked one up it was hot, like it just came out of the oven. I was out back hanging clothes when I saw them come tumbling down the driveway and hit up against the truck and bounce off in all directions. After we'd go to bed at night it stopped. Then the banging on the walls started. It kept up till bedtime, too. These things went on until right around the time the shack came down." "After grandma, Tommy kept playing in the shack, digging up things. I didn't stop him; I didn't know it was a warning for him to keep out. Then one day, down it came like a load of lumber. Lucky for Tommy he wasn't in there."

Andrew: "What kind of a structure was it?"

Janet: "Wooden ... old rotted boards with a corrugated roof. It always leaned with the wind; but I thought it was sturdy enough. I guess it was a coop, or a storage shed ... something like that at one time. When it collapsed the lumber was piled up fifteen feet high. I know if he'd a been in there, he'd a been killed."

Andrew: "When did this *take* place?"

Janet: "Just a few days ago."

Andrew: "Has anything happened since ... anything strange you can't account for?"

Janet: "No. Not since then."

Looking at the case from a supernatural viewpoint, it could be argued that the family had been forewarned of impending danger. It started when "great-grandma" announced her arrival by shaking the truck, by raining rocks on the roof, and banging away at the walls. Next, she led Tommy to the tin of snapshots. As a recognizable spirit, she told him to 'Beware.' That the McDonalds failed to heed her repeated, albeit obscure warnings, did not make great-grandma a bad guardian. Tommy had, after all, escaped injury.

There's another explanation, a psychological one related to the emotional condition of Janet McDonald. At the risk of sounding chauvinistic, we know that women without a man or other adult companion in the house (even today's self-sufficient women) do seem to be more prone to experiencing psychic phenomena. Some, apparently, express the fear of being helpless to defend themselves through subconscious projection of threatening noises and shadowy figures. Psychologists know that divorcees are apt to be emotionally unstable in the early stages of separation. Objectifications of PK and ghostly forms may be the result of a subliminal desire not to be alone. As far as I am aware, no statistics are available on men living alone, but I'm sure they apply equally.

In addition, we know there's an atmosphere of extreme tension present in children of divorce. A volatile situation can become explosive when one of them is approaching puberty. As fate would have it, both Tommy and Jessica had reached, or were about to reach puberty. In On *The Track Of The Poltergeist,* the late D. Scott Rogo warned us, "Some cases that look like traditional hauntings are nothing of the sort. They are actually disguised poltergeists being created by the people currently living in the house." Substitute Directed RSPK for "disguised poltergeists," and Scott could have been describing this case. Accounts of rocks mysteriously raining down from the sky are commonplace to newspaper reporters and Fortean scholars. Yet no one had offered a reasonable explanation for them until Professor William G. Roll labeled them RSPK. There are, of course, incidents in which children are responsible for throwing rocks; but I don't think it applies in this case.

Because the locus of these attacks is often the home of a pubescent youngster, researchers have identified "stonings" with the mischievous Poltergeist. It is interesting, as well as typical of RSPK, that the rock fall and the banging sounds stopped as soon as Janet and the kids went to bed.

I'm convinced that the McDonald's apparition was Directed Adolescent RSPK and not a visitor from the nether world. Sometimes late in a RSPK haunting the Agent's subconscious mind projects a glowing life-like figure. If that's what took place, the warning to 'Beware' could have been a subliminal attempt at self-preservation prompted by intuition. It may be stretching it a bit, but in that case both the photograph and the collapsing shack were probably nothing more than impressive displays of Jung's synchronicity.

CASE FILE: THE WAILING

"Not louder shrieks to pitying heaven are cast, When husbands, or when lap-dogs breathe their last." (Alexander Pope, *The Rape of the Lock)*

One of my cases reinforced my conviction that what we are exposed to every day in life is only a small part of the overall picture. The following was transcribed from the tapes of February 21, 1991:

"It's 7:20 p.m. Andrew Nichols is in the home of Elizabeth Barnett. Gathered here, in addition to Mrs. Barnett, is her son Sean and his wife Darlene; daughter Vickie and son-in-law Norman, and their boy Johnny; and Elizabeth's friends, Terrence and Judy Frazier. (All the participants' names have been disguised at their request). "The purpose of our visit is to conduct an investigation with the aim of determining the source of a high-pitched wailing sound":
Andrew: "Would you fill me in on what is going on here, Mrs. Barnett?"

Elizabeth: "I have noticed noises in the house for quite a few years now."

Andrew: "Your son (Sean) told me that it started the year he and Darlene got married ... the same year he moved out of the house."

Elizabeth: "That's right. It was March 1977. Gerald, my second husband, and I...my first husband, Sean, Sr., died in 1969 ...Gerald and I used to hear noises almost constantly. And he would say, Well, perhaps this is Sean and Lilly calling to us.' Lilly, of course, was Gerald's first wife. She passed away shortly after Sean Sr.. We were all very good friends. And he and I would say, 'Maybe they've gotten together over there to get back at us for getting married.' And the noises were so pronounced at times and so loud that we thought, *Well, they're just unhappy with us ...for what reason we don't know.* And we'd dismiss it."

"I told the children about it and they said, 'Oh, sure mom! Uh-huh! Well, my daughter Vickie's birthday was last December fifth. We all went out to dinner and came back here for cake and ice cream. That evening I had Christmas records going in the family room. We were in the dining room, and I had the record player up so we could all hear it. All of a sudden it got quiet, and then it started. Gerald was there (although she and Gerald were divorced the year before they were still the best of friends), Sean, Darlene, Vickie and Norman, Terrence and Judy. We were all there and, oh yes, Johnny, my grandson. "They became so pronounced that we all looked at each other, you know. And my daughter-in-law said, 'Is *that* what you've been hearing?' And Gerald and I nodded our heads, 'Yes.'"

Andrew: "All of you could hear it?"

Elizabeth: "Yes, indeed."

Andrew: "Did you compare notes on what you heard?"

Elizabeth: "Yes."

Andrew: "Do you all think you heard the same thing?"

Vickie: "Yes. But one interesting thing ... we all said it came from a different place. One said it came from the attic; one said from the basement. I said it was right there in the dining room. Now maybe it depended on where you were sitting. Sean turned the stereo off completely; but we still heard it. And my little boy (she reached around Johnnie and gave him a hug, he got under the dining room table. He was so frightened he wouldn't come out."

Johnny: (Mimicking what he'd heard) "Aaarrrh. Aaarrrk "

For some reason the idea of Johnny hiding under the table made me think of a scared puppy.

Andrew: "Do you have any animals in the house?"
Elizabeth: "No."

Andrew: "Have you ever had any pets in this house?"

Elizabeth: "Well, we had them as boarders ... my children's pets when they'd go on vacation."

Andrew: "But no household pets."

Elizabeth: "No ...I don't have them."

Vickie: "Tippie!"

Elizabeth: "Well, Tippie has been gone for years."

Vickie: "One interesting thing ... this started after you got rid of Tippie."

Elizabeth: "Tippie was a bad dog."

Sean: "Tippie was a poodle ... a little black poodle."

Darlene: "We never thought of that!"

Elizabeth: (Obviously amused by the idea) "No, we never did. We had to put him to sleep."

Andrew: "You put him to sleep?"

Elizabeth: "Uh-huh. We had to. After Sean (Sr.) moved out ... he'd gotten so out of hand."

Andrew: "Let's go back to the sound you heard. Can you describe it for me?"

Elizabeth: "Uh-huh. Woo-a-woo-a-WOO-a-wooooo."

Andrew: "Was it that high-pitched?"

Sean: "No, it was deeper."

Elizabeth: "No ... higher. Well, that's as high as I can go."

Terrence: "I don't know, it sounded more...."

Andrew: "If you had to guess whether it was human or animal what would you say?"

Judy: "Human."

Darlene: "Oh yes ... definitely."

Sean: "No ... animal."

Terrence: "Animal!"

Elizabeth: "Human *and* animal."

Andrew: "Male or female?"

(In unison they shouted "male!")

Vickie: "What it sounded like to me was when our father, Sean, Sr., was in the hospital and there was massive brain damage. And they would go in and suction; and he would moan and grown ... like someone in pain and misery."

Terrence: "It was just a moan ... three of them in succession."

Judy: "And they got louder."

Andrew: "The sounds that you've heard, Liz, were they all similar? If it were on tape, would you think it was the same tape played over and over again?"

Elizabeth: "Yes. Yes. Only louder at the birthday party."

Andrew: "It's getting louder?"

Elizabeth: "It was that night. The intensity that night was very strong."

Andrew: "Have you heard it since then?"

Elizabeth: "Yes. Last night. After I went to bed, it was around eleven-forty, it began. I stood in the dining room with all the lights out. I always look out the dining room window to see if there's a wind blowing. There was, but I don't see how it could have made the sound."

Andrew: "I don't want to insult any of you ... obviously you're all intelligent people ... but you don't have any kind of air vent on the roof that..."

Sean: "No. It's a sealed roof...all closed in. We don't have the spinner type vents, if that's what you mean."

Judy: "Elizabeth, didn't you say you've heard it going on when there was no wind outdoors?"

Elizabeth: "Yes, often ... often."

Judy: " And you even checked whether it had to do with the furnace."

Elizabeth: "We turned the furnace off; it didn't help any."
Terrence: "If it was a man, I've ... I've never heard anything like it before."

Andrew: "Is it the kind of sound that any of you could produce without too much difficulty?"

Elizabeth: "Well, I tried to do it, but Sean said it was too high."

Sean: "It was just like an Aaarrrh ... a long moan."

Elizabeth: "Sometimes they're jabs."

Judy: "Yeah. Short jabs."

Darlene: "And it increased in intensity and got very loud and then stopped and started again ... three times."

Judy: "Short jabs ... and it was over a span of maybe six or seven minutes."

Andrew: "Okay. Is there a possibility that each time the sounds have been heard there was a television set, a radio, or a record player on?"

Elizabeth: "No. Unh-unh. Except for that one time it's always been quiet when I hear them."

Sean: "I remember, for example (turning to his wife) ... when was it that we came over?"

Darlene: "The night we took her out for her birthday."

Sean: "Oh yeah! We walked in the door and she had her coat on ready to go ... that was the anniversary of my father's death, January the third, and the three *of us* were going out to dinner. And she said, 'Listen."

Andrew: "You heard it, too, that night?"

Sean: "Oh yeah. I've heard it several times."

Andrew: "Do you always hear it at the same time of day or night?"

Elizabeth: "At random times."

Andrew: "Did both you and Gerald agree that it was a male voice you heard?"

Elizabeth: "Oh no. When we were living together I used to say, 'It's Sean; and he would say, when he heard it, 'It's Lilly.'"

Andrew: "Has anything else of interest happened?"

Elizabeth: "Yes, it has. Last night ... after I stood in the dining room listening to the moaning...it stopped, so I said, Okay, I'll go to bed. I went in and read for a while... Right where the crucifix is and where the rosary hangs on my bedroom wall, I heard ...I swear, it sounded like two shots. Now I know how wood sounds when it contracts ... these were two definite BANG! BANGS! So much so, I sat up and thought, *I never heard a noise like that before.* I never heard two cracks like that in my life before. "Just before Judy's birthday, which was the fifth of December, I was talking to Sean on the phone and all of a sudden I was aware of footsteps. They sounded like they were emanating from the pantry door to the end of the counter where my telephone is ..... and I didn't want Sean to know I was frightened. I thought, *I have a burglar alarm; I came in and locked the system. Okay, nobody is in the house. But why do I hear these footsteps?* They sounded like they came from the pantry and stopped right along side of me. I hung up and went through the whole house ... the basement and here; I can't get up into the attic. I heard invisible footsteps, but didn't *see* a thing."

Andrew: "It is due to the *number* of phenomena that you want help with this thing?"

Elizabeth: "I'm not frightened. When it started we just thought that, *Well, if they're* (Walter, Sr. and Lilly) *with us they just want to be with us; they want to be near us."*

Andrew: "Do you feel more concerned about it now?"

Elizabeth: "Yes, because I'm alone now. Last night was the first time I was truly frightened. The footsteps and the bangs last night ...I was frightened. Sean has provided a burglar alarm and a garage door opener. I was feeling perfectly safe in here alone; but I was frightened on those two occasions."

Andrew: "None of you could recognize the moaning sounds?"

Sean: "Maybe it's more a groaning sound than a moaning one."

Elizabeth: "It changes. It isn't always the same."

Sean: "I've never heard a noise like that. I've never heard a human being make a noise quite that loud. It'll make the hair on your ..... on your arms, or wherever, stand on end (no doubt paraphrased in deference to my balding head)."

Darlene: "I've never heard anything like it in my life."

Vickie: "We never believed it. We'd just make fun of mom when she'd tell us."

Elizabeth: "And 1 have to put up with it every night!"

Vickie: "You know each of us has spent a night here alone when mom was traveling- Sean and Darlene, and Norman and me, and we've never heard it."

Andrew: "You mean you've never heard it when your mother was absent from the house?"

Vickie: "Yeah."

Sean: "Has anyone heard it when my mother wasn't here? The answer to that is no!"

Judy: "Elizabeth, didn't Gerald used to hear it when you used to work?"

Elizabeth: "No. I don't think so. I'll have to ask him." (Later, I learned that he hadn't.)

Andrew: "Do you have an interest in the occult, in ghosts, flying saucers, and things like that?"

Elizabeth: "No. Unh-unh."

Judy: "Didn't your grandmother, or someone have..."

Elizabeth: "Yes. My grandmother said she was born with the 'caul.' That's an Irish term ... and she was *born* in Ireland. She claimed she could foretell the future. I hope I haven't inherited her talents."

Andrew: "Did you ever try to talk with the person you think it might be while he was moaning?"

Elizabeth: "Yes. I said, 'Okay Sean ...I guess it is you. If you're still in purgatory I'll say a prayer for you. I hope you're happy now ... may you rest in peace. Manifest yourself. .. show me what you're doing. Show me yourself. Give me some signal.'"

Andrew: "Was there a change in the moaning at the time you said those things?"

Elizabeth: "No. I just talk to the moaning. I say, 'Now, c'mon Sean ... you left me ... you left me too soon.'"

Vickie: "You blame him, mom, for leaving you alone! You haven't forgiven him."

Elizabeth: "No (softly)."

Vickie: "You know you said that recently..."

Elizabeth: (With eyes lowered) "Maybe I have said that. I have ...I know I have."

Andrew: 'This is something you've done a number of times recently?"

Vickie: "Last summer, at the time of her divorce, was a very hard time for her."

Elizabeth: "I guess. I had problems. Sean was gone; I'd entered into another marriage that didn't work. That's why I always say, 'You know the condition I'm in Sean. .. now help me!' I talk directly to him; I'm sure he can hear me."

Sean: "I remember, mom ... this must have been three weeks ago. Remember, I ran downstairs (to the basement), and it was really going. There was Vickie, mom, and myself. We were at the top of the stairs, and it was really going strong; so I ran downstairs and it stopped. I got down there and I looked all around. And, I remember, I got up stairs and it cut loose again ... you know, like I dared it and so it came back."

Elizabeth: "And I don't want to dare it ... not if it could be demonic. That's what I don't want. I have to live here by myself after the rest of you are gone."

Andrew: "I don't think you need to be concerned about that."

Vickie: "The night of my birthday we were very loud. And the stereo was loud and we could hear it above all that. We never pushed mom to do anything about it because ...(looking around at the others) frankly, I never heard it before."

Elizabeth: "Sean loved us all, but Vickie was his favorite. It figures he'd show up for her birthday."

Judy: "You know, I believe in this."

Sean: "Well, sure. You have to believe in what your own senses tell you."

Andrew: "And yet, it may be nothing supernatural at all. It may have nothing whatsoever to do with the dead, or the demonic."

Sean: "A couple of hundred yards back is an old German cemetery. There was never any indication that the grave sites came this far down, though."

Sean had suggested the theme of Steven Spielberg's *Poltergeist*. I closed the interview with my customary advice:

Andrew: "The more attention you give this thing the more it'll grow. You, Mrs. Barnett, are obviously the focus of these sounds. If you can turn your attention away from what's happening, more than likely it'll stop on its own. If not, and if you want me to try to remove the source, I'll schedule another visit. You do want to be rid of it-or do you? Am I assuming too much?"

Elizabeth: "No. I would like to understand what it is. I wouldn't want whatever it is to be banished to punishment, or anything. But definitely I'd like to get rid of it."

Andrew: "I wouldn't be banishing it in the sense that I have powers to send it to purgatory or hell. I'd be blocking it from making a nuisance of itself."

Terrence: "You don't think ...I mean, what's your opinion about it being a relative ... do you feel it is?"

Andrew: "No, I really don't."

And I didn't. The case had, on the surface at least, many of the characteristics of Random RSPK, and none of a discarnate entity. Circumstances dictated that I return to Elizabeth Barnett's home (February 30, 1991). The family brought me up to date:

Sean: "Mom tells me they seem to be amplified throughout the house."
Elizabeth: "They come mostly at night, now. I think they've reached a peak, not only in volume but in activity. I've heard them approximately a dozen different times since your last visit."

Andrew: "What do you do when you hear them?"

Elizabeth: "It's usually after I'm in bed. I get up with the lights off. .. walk slowly into the dining room, sit down and listen."

Andrew: "Are you still trying to make contact with him ... or it?"

Elizabeth: "Yes. I say, 'C'mon Sean. Help me out of this situation you've left me in by dying. Let me see you. Give me a sign.'"

Andrew: "In other words you would *like* to communicate with him?"

Elizabeth: "Yes, I would. I say, 'Help me out of this mess you've left me in; the children have problems; Gerald and I have troubles.' (He would like to get back together, but I don't think that's a good idea.) 'Give me a sign so I'll know what to do.'"

Andrew: "Is there some reason you think he can help you just because he's on the other side?"

Elizabeth: "Yes, because he always helped me. He was always there, even before I asked for help he'd offer it. He knows it's been unbearable for me without him ... even after all this time."

Andrew: "Did you have some special reason for believing you could contact him?"

Elizabeth: "I'm sure he can hear me. He's probably listening to us now. I helped him find God before he died, so I threatened him. I said, 'C'mon Sean. I helped you get where you are-the least you could do for me is to help me now.'"

Andrew: "There is a theory, you know, that you can hold a spirit earthbound by summoning him up too often. Even if it isn't a supernatural spirit, your insistence that Sean help you may be producing the energy needed for whatever this thing is to be heard."

Andrew: "You haven't said, do you take any kind of medication?"

Elizabeth: "Yes ...Inderal and Libertab for my heart."

Andrew: "Sometimes people who take hypertension medications are subject to strange noises. It's in their 'mind's ear' so to speak. But that would pertain only to that individual and not to..."

Elizabeth: "Others (she finished my sentence). Others couldn't hear it."

Andrew: "Well, they might hear it telepathically if there was an emotional content. It's possible you may need this contact emotionally. You keep saying you wish he was here."
Elizabeth: "I don't need *Walter,* I just need his help! I think you're emphasizing my needs too much. I'll never forget him, but if he were sitting there now I'd probably badger him to death."

Vickie: "But you're *always* calling out to him. You make it sound as if it was his fault for leaving you. I'm sure if he'd have had a choice he wouldn't have left us."

Elizabeth: "I know."

Andrew: "The best thing is to pay more attention to this world. You have the ability yourself to solve your problems. These sounds would be gone tomorrow if you wanted them to be gone. Our subconscious tries to please us. It has the ability to create the sights and sounds needed to please our conscious mind."

Elizabeth: "Why don't I hear these noises when I go elsewhere? You know, Sean was never in this house. He died before I moved here."

Andrew: "Most of this kind of activity is confined to one place. In most of our cases phenomena run their course then end. I think you keep reinforcing these noises by demanding that he come back to help you; so it just keeps keeping on."

Terrence: "Ah! surely *nothing* dies but *something* mourns! (quoting Byron's *Don Juan;* then pausing briefly before moving on). It could be, Elizabeth, that you made Sean, Jr. his father's replacement when he died ...I mean as far as responsibility for your welfare is concerned. Then when junior got married and left it caused a psychological crisis that created the noises you now attribute to Sean, Sr."

Elizabeth: "But you can't say that a four-year old child who isn't aware of the conscious or the subconscious could imagine something like that. Normie actually heard something that scared him."

Andrew: "It's a hard concept to follow, I think. It is possible that you are projecting these sounds ...I mean literally making them, subliminally ... yet everyone around you is able to hear them too. The only way we could prove it, one way or the other, is to get them on tape. If they can be recorded they exist. You can't tape record a hallucination. That doesn't mean it's supernatural if it's taped, only that it has substance. The source, we believe ... if it isn't some natural occurrence ... is still your mind. Your subconscious mind will have done it psychokinetically. The chances are that it *can't* be recorded; that it is telepathically transmitted to the others."

Sean: "Why do you say that?"

Andrew: "There are so few genuine recordings of ghosts on film or on sound tape. Once taped, though, we can take it to a sound lab and have it analyzed. It could be anything: an airplane, ultrasonic waves, or even the roaring traffic of the Interstate (the house was only a mile or two from I-40). But if you have it on tape they can track it down. I'll tell you one thing ... it's nothing supernatural, so you don't have to worry about that. It's nothing that's malignant."

I summed up my second visit in my notes:

> "After reviewing the facts I arrived at the conclusion that Elizabeth Barnett is subconsciously producing the moaning and howling sounds as a fulfillment of her desire to have her deceased husband back. She may be making Sean, Sr. suffer--figuratively speaking--for leaving her prematurely. Her constant calling to him for help and her condemnation for leaving may be creating the sounds. Witnesses seem credible. If so, they're hearing it telepathically or by way of psychokinesis. The tape recorder should decide that issue. I suggested that Elizabeth stop blaming Sean, Sr. for dying and stop calling to him for help. If she did, the sounds would probably cease".

Possible explanations:

- Misinterpretation of a normal event
- Shared hallucination produced through autosuggestion
- Random Adult RSPK
- Fraud, with or without collusion
- A prank played on the family: a loud speaker in the duct system,

etc.

No experiments were undertaken or planned for a future date.

After not hearing a peep from the family in six months, I got a phone call from Sean Barnett, Jr. His mother had been in Florida for the past two weeks and was due back the next day. Sean and his wife Darlene had--he was embarrassed to admit--put off checking the house in her absence until the last minute:

Walter: "Yesterday, as soon as we walked in the front door, we heard it ... Darlene, too. It was groaning so loud I thought mom had left the television on full blast. We stayed long enough to water the plants, then got the H out of there. Since you were here she's only mentioned it once or twice; and that was in the first few days. Trying to be consistent with what you told us, I said, 'Ignore it.'"

I suggested that Sean refrain from telling his mother about this latest incident. I was afraid the cycle would start again if she knew. With her hundreds of miles away it was a little surprising that they could manifest --unless, perhaps, they were something other than psychologically generated effects. Of course, I knew that Random RSPK (specifically the Poltergeist) could be heard-from when the Agent was away from home. I was soon to learn that Elizabeth had endured far more disturbances in the intervening period than she was willing to admit to her son. On the night of August 9, 1991, I received a call from a badly shaken woman. The moaning had been nearly unbearable since she returned from Florida:

Andrew: "So they've started again. I was afraid of that."

Elizabeth: "They've never stopped ... not really. I tried to tell Sean, Jr., but I don't think he wanted to hear it. I wasn't going to call, but who else do I have? There was no sign of them, not once, all the time I was away. It seems as though evenings, especially around midnight are the most frequent times. I was terrified last night because they started in their usual high-pitched howling and then decreased into a low, manlike moaning tone. I could almost make out words this time."
Again, I assured her that she was in no danger. I asked that she call me, even in the middle of the night, if it started up. I paused for a moment, then resumed:

Andrew: "I'd like to try to pick it up it over the telephone, if I can."

Elizabeth: "I've got it on tape! Or, I should say, I *think* I have it. I haven't had the nerve to listen, so I'm not positive."

I was at her house the next day to pick up the cassette when she told me in a low, fascinated voice: "On July the twenty-ninth, I heard another loud 'shot.' It came from the vicinity of the crucifix and rosary again. Now the moaning seems to be mainly in the dining room. I can't be absolutely certain of it, since it's so sporadic and inconsistent, but it seems to be coming from there. "I am absolutely certain that I have not been responsible for these sounds. I have been religious about not asking my late husband for help. I say, 'I can take care of myself. You go on about your business and don't worry about me.'"

That evening I listened to the tape. Only side "A" had anything on it. Although barely audible on my portable, the sounds were undeniably there. When I switched to stereo they became chillingly real. Vickie was in the basement when the speakers began to put forth their prolonged mournful cry. In a flash, she was up the steps demanding, 'What in the world is that?' “That,” I replied, “is possibly not of this world.” It wasn't melodramatics. I'd been knocked off my pins. Someone had opened the gates of hell and the "howls of the damned" were reverberating throughout the house.

It could be argued, I suppose, that suggestion and imagination had gotten the better of me. I recognize the effect these twin influences have on our balance. Yet most who listen to the tape, as well as to those subsequently recorded by Mrs. Barnett (three full hours of them), have reacted the same way. They're held in its grip, describing the sounds as they rise and fall as a demented woman crying; the shrieks of the tortured; or the shrill cries of the banshee. Others say they're a combination of wailing wind, wounded animal, and whimpering child. None have suggested how such diverse noises might have been produced.

Only once had I heard anything even remotely similar. It was in *ALIEN,* the 20th Century Fox hit movie of 1979-80. It blasted eerily onto the scene when the crew of the space barge arrived on the mystery planet (just before running into the monster pods). The mechanically reproduced sound of wind is disturbingly similar to the Barnett's's howling. So much so that I wondered if someone had been playing it for their listening displeasure: piping it through the duct work, perhaps.

Later, when I mentioned my suspicions to Sean, Jr., he told me that they were viable, but he had already paid a small fortune to an acoustical engineer to check it out. Natural sounds, within or without the house, had been eliminated as a factor. Not even the roar of traffic from the Interstate could account for them. "He heard it for himself," Sean added. "Even measured it with his instruments; but had no idea what it might be."

I heard from Elizabeth on the twenty-second and twenty-third of August. On each of the previous nights she had resisted the growing panic in her mind, crawled into the dining room in the dark, sat on the floor and transcribed the lamenting wails of her nocturnal visitor; then called me so that I might hear them, too. One description--that they resembled the cries of the banshee--was particularly interesting considering her and her late husband's Celtic heritage. In Irish lore, the term means "fairy woman," but a banshee may be an ancestral ghost bearing somber tidings. Before the death of any member of her clan, she's heard weeping and wailing for days.

An earwitness to this forlorn foreteller of doom, when describing what she'd heard, said: "It started low at first, then it mounted up into a crescendo; there was definitely some human element in the voice ... and you could almost make out one or two Gaelic words in it; then gradually it went away slowly." (Excerpts from *The Supernatural: Ghosts and Poltergeists*, Aldus Books, Ltd. London.) In 1937, Elizabeth O'Malley had married the Barnett boy, Sean; the family was as Irish as Irish could be. But Knoxville, Tennessee wasn't the "Old Sod." And even if it was, no member of the O'Malley or Barnett clans had passed on between the death of Sean, Sr. and the start of the "keening" (a wailing lament for departing souls). Clearly, if it was a banshee, it was a mite confused.

So I returned on Sunday night, August 24, 1991. This time we brought along a young psychic sensitive named Susan Garrison. Susan, a graduate student at the University of Tennessee, came highly recommended as a talented novice medium. The endorsement of a medium was no small matter. Self-proclaimed psychics were at every meeting of the local Psychical Research Society. There were women (and some men, too) who would attempt strange oracular feats: like "interpreting" the yellow streaks in your handkerchief; then suddenly backing off, refusing to disclose the "truth" they'd seen in the nose effluence. "You're better off not knowing these things," they'd say. I were determined not to subject our clients to anything so bizarre. Susan's credentials had to be impeccable, and they were.

Frankly, I was beginning to lose faith in my original diagnosis: that Mrs. Barnett was subconsciously creating the disturbance (probably the natural outcome of hearing that god-awful thing myself). Although this curiosity seemed to be more the result of the fulfillment of some psychological or physiological need than the manifestation of the supernatural, the possibility still existed--however remote--that an other world agency was involved. If we failed to reach this thus far incoherent source using "normal" means, I had hoped Ms. Garrison could tap into it psychically. If not, there was the chance that she could at least eliminate Sean, Sr. as its cause.

At 7:30 p.m., I introduced Susan to Elizabeth, her son Sean, and Gerald Mancini, Elizabeth's ex-husband. Only six of us would take part in the experiment: "Ms. Garrison is a psychic sensitive who works with us from time to time," I began. "I suggest that we not discuss what has already taken place in front of her. Susan's been told nothing of the details of the case beforehand; her impressions should be free of preconceived ideas."

We dimmed the chandelier, then sat down at the dining room table. Elizabeth, Sean, and Gerald operated the writing planchette; the rest of us leaned forward to watch. Responses to my questions came in quick, furious movements of the tripod. None of them, unhappily, had anything whatsoever to do with the case. We continued until just after nine--getting nowhere--when I uncharacteristically lost my cool and blurted: "We have asked that you help us in our attempt to help you break the restraints that are keeping you earthbound. If there is any way within our power to help you find your rest we will do so. But this we must tell you: if you remain here against the wishes of this family, your immortal soul is in grievous danger. We command you to leave this house at once and go directly to your appointed place in the other world, there to remain forever!" These were no off-the-cuff threats. They were "ghost-be-gone" words used in the layman's exorcism. Not surprisingly, everything came to a halt. The planchette fell limp, and Elizabeth stifled a gasp--fearing, no doubt, that I'd undone her efforts to save Sean, Sr. from eternal damnation.

At that moment, something was going on outside the dining room window. We all heard it, but Gerald was the first to spot it. Seated facing the window, he saw the flower laden spirea bushes beat against the pane. We all watched while, like palm trees in mid-hurricane, they thrashed around and slapped against the front of the house. Susan and I raced to the front door. In the excitement, I felt something low to the floor bump up against my leg: *imagination,* flashed through my mind.

Once outside, we saw that the bushes were the only things moving on an otherwise calm summer night. White petals were strewn everywhere. Suddenly, they stopped their frenzied dance. From midway up the nearest clump of elder trees came a high pitched shrill. It was soft but unmistakably the cries on the tape. The two of us stood there listening for several minutes, mesmerized; then started back to the house. As we approached the door, I said: "You did hear it, didn't you?" "Like a screeching animal?" "Yeah. Did you sense anything as we came out the door?"

"Wait ... let me jot down my impressions," she said. "I'll tell you later." We made our report to the others. Elizabeth was dumbfounded: "What was my *husband* doing up in a tree of all places?"

The seance reconvened. Now Susan was operating the writing planchette alone:

Susan: "Why have you been here so long? How can we help you? Who are you ... what are you?"
(There were no answers.)

Elizabeth: (Placing her fingertips on the tripod near Susan's) "If you are unhappy, Sean, please let me know. How can I help you? I'm not afraid of you ...I know you won't hurt me. C'mon darling. I think I know who you are, and I want to help. Please talk to me." (Nothing.)

Sean: "Ask it to move the bushes now ... if it's outside."

Elizabeth: "If you are there move the bushes outside the dining room window." (Nothing happened.)
Sean, Jr. and Gerald added their hands to the tripod; it began to initiate wide sweeping circles:

Andrew: "That's a good sign."

Elizabeth: "You know Gerald and I love you, and we'd never do anything to make you unhappy. Why are you *doing* this to me?"

Andrew: "Ask it again, Who are you?"

Elizabeth: "Please tell me who you are."

Susan: "Come back in the house. Please come in from outside."

Elizabeth: "Are you in physical pain? You shouldn't be. You should be at peace now."

Susan: "Why won't you identify yourself?"

I had made a mistake by threatening it, I acknowledged to myself.

At ten-thirty we took a break. Susan and I walked into the foyer: "I think it's strong and angry," she said with conviction, "but I'm not sure we're going to get any kind of message. Here, I wanted you to see this." Earlier, while touring the house, Susan had written a few thoughts and impressions in a small notebook: Dining room: chills, sadness, crying, strong feelings front of window. It walks away and dies down. Bedroom (Elizabeth's): screaming; hysterical crying; hatred; unhappiness. Basement: Fear, hyperventilation; fast, short panting breaths; something bustling through, or around; a strong presence sends chills through me; hair on end; move around freely; have to get out of here to sky and clouds and wind rushing. To these she had added: On way outside: Something in foyer, black and furious rushing along side, brushes against leg. (I, too, had felt something bump into me in the foyer.) Front yard: strong presence just outside the dining room window; something in tree crying. An animal? Suddenly, we were interrupted by short, loud shrieks. I glanced at my wristwatch; it was ten thirty-eight. The "gate" had swung open!

Elizabeth: (Running into the foyer) "Do you hear it? Do you?"

Andrew: "Yes. Of course."

Elizabeth: "Now this is the first time anyone but immediate family and Terrence and Judy has heard it ... isn't it?"

Gerald: "It's like a siren until it gets in voice."

Sean: "Yeah."

Susan: *"Shhh!* Let's listen."

Elizabeth: (Pulling me aside) "I'm glad you got to hear it."

Andrew: "Is it as loud in here as it is in the bathroom?"

Susan: "Well, listen!"

Susan and I raced through the sprawling ranch. Everywhere--from all directions at once--the sound of desperate wailing could be heard. The hairs on the nape of my neck stiffened in response to it. It was dreamlike; the sorrowful yowling cries of a wide awake nightmare. From the basement, Walter called out, "It's down here. Wait a minute. No, now it's gone." After ten minutes of whooping it up, it'd quit: stopped just as mysteriously as it had started. Mentally, I applauded. It's not often I'm entertained by my clients' oppressors, nor rewarded for my near-fanatical persistence.

Susan: "My feeling is that it's a very angry animal ... a black cat, or a dog; maybe a Doberman, or something smaller. It's smart, but mean and vicious; and even though it may not always sound like an animal, believe me it is!"

Andrew: "Didn't you mention a black dog when I was here in February?"

Elizabeth: "Tippie, our poodle."

Gerald: "That was one mean little dog."

Elizabeth: "I just never thought..."

Sean: "You put him away right after I got married"
Gerald: "I want you to see my hands ... the chunks he took out of them. When you say that little bastard was vicious, you're telling it right."

I counted half a dozen scarred-over notches where Gerald claimed the dog had bitten him.

Gerald: "Show 'em your battle wounds, hon."

She had them, all right. There were light patches of scar tissue on both of Elizabeth's hands and several on her right leg, below the knee.

Andrew: "How could you let that happen?" (What I was thinking was *How could you let it happen more than once ?*)

Elizabeth: "Poor Tippie. He wasn't always that way. Only at the end did he get so feisty. We'd lock him up in the bathroom ... the one in the basement, when we left the house. He'd tear things up if we didn't."

Gerald: "That's what led to his downfall. One day we came in from the garage, and there was Tippie defiantly standing there barking at us and baring his teeth."

Elizabeth: "He'd escaped from the locked room."

Gerald: "Chewed and clawed his way right through the wallboard under the sink. Ate a whole right through the wall, and then just stood there in the doorway, defying us to lock him up again."

Elizabeth: "That's all Gerald could stand. Off to the pound he went. I guess they put him away; we never really knew. But if it is Tippie, how in the world did he get up a tree?"

Human beings have a penchant for drawing erroneous conclusions from what they observe. Even parapsychologists get chinks in their equilibrium. Besides the tape recording, we had four years of anecdotal material and ten minutes of evidence--all of which was merely suggestive, not proof. Yet this doleful dog, if that's what it was, had begun to get to me. When Susan Garrison suggested that Neptune had returned with his "Irish up," I had to admit--except for his trip up a tree--the explanation was as good as any.

But how do you get through to a vengeful animal? I don't do dogs, so I suggested that Elizabeth ask her parish priest to bless the house. I also suggested that she stop feeling remorse over sending "poor Tippie" to an early grave. Feelings of guilt, I reasoned, may have created the sounds of the suffering pet. Listening to it night after night could be the fulfillment of a subconscious desire for punishment. There was also the possibility that, in the end, the creature that bit and tore its way out of confinement was no longer Neptune.

I hadn't given the "dog in the elder" episode much thought until an article, "The Mystic Powers of Trees," appeared in the February 1990 issue of *FATE*. In it, author Louise Riotte traces the ancient history of tree worship and mankind's relationship to these majestic sources of energy. She speaks of special groves in Greece: "...where people could go to worship and give thanks to the trees and the spirits believed to reside in them. Most trees give off wonderful healing energies," she tells her readers, "but even so, there are some that are malignant and dangerous." Trees, according to the author, are nature's exorcists ...... taking spirits--some of which may be very dangerous to human beings--into themselves, their branches and bodies." Tradition suggests that they may also keep bad spirits trapped safely in the ground where they can do no harm. "A living tree was believed to be an easy focus to receive spirits because it holds them effectively. Even today, some people believe that a good exorcist can send a harmful entity into a tree if he is unable to cope with it in any other way."

Elizabeth Barnett mentioned that several disease-ridden trees in her yard had been felled late in 1986. Were they imprisoning a spirit who, once released, searched for a lower form of life to inhabit? Was a malignant force possessing Tippie at the time of the little dog's death? And was the wailing only a continuation of this force's persistent harassment of our client? My instruction to it to go directly to its appointed place may have sent the evil thing back--if only temporarily--to the branches of the elder tree. Our pleas for its return may have coaxed it into the house again.

The taped reproductions, roughly two hours of unearthly sounds, are among the few pieces of hard evidence I have assembled in over twenty-five years of collecting memorabilia. Whatever the force responsible, I am more impressed by these pitiful noises than I have words to describe. To me the wailing is an ill-wind come alive; it is suffering borne on a howling current. It is piercing grief in a series of long-winded moans; each sustained for a length of time far in excess of that which any air breathing organism could match; then falling off into a series of short sobs and whimpers. How could Elizabeth Barnett share her home night after night with that bloodcurdling cry? How could she do it alone?

Eventually the sounds were analyzed by two different labs. The results were dissimilar. One acoustical engineering firm said they were quite unlike normal noises. Shown on an oscilloscope, ordinary sounds have a distinctive curve, rising and falling in slope-like fashion. These, according to the first set of experts, began and ended abruptly as though they'd been artificially manufactured. The second testers, technicians who use a spectrum analyzer in their work, disagreed. What they discovered was a strong signal, but one that was composed entirely of natural (human or animal) sounds.

I believe that natural noises (the wind, for example) and mechanical, or electrical vibrations may be psychokinetically transformed by the PK Agent into words; into singing; laughing; and even wailing and whimpering: a process akin to shaping the twanging tones of a jews' harp. In this case, however, no sign of any contributory vibrations could be detected by either set of experts. "It's a one in a million shot," said one critic--assuming the sounds were an atypical fluke. It's difficult to deal with those who say haunting phenomena are phony because the odds against them repeating are astronomical. When I found myself listening to eerie sounds that came at me from every direction, I didn't ask, "What are the odds of that happening again?" The odds of it happening at all were astronomical.

Early in November 1991, I got a phone call from Elizabeth. Her priest had been to the house twice to bestow the blessings of the church. The wailing diminished to a degree with his first try. The second attempt had chased it completely. There is no taped record of the conversation that followed, but I've managed to reconstruct it from my imperfect memory:

Elizabeth: "I almost called you to ask if you'd come to our Halloween party. I'm glad I didn't. Something terrible happened. We held a seance here that night. I know ... it was a foolish thing to do, but we did it. One of my guests, a good friend, collapsed and died."

Andrew: "Died there ... in your house?"
Elizabeth: "No. No. It happened later that night, after he and his wife left the party."

Andrew: "Oh, I don't think there's a connection."

Elizabeth: "You think not? I hope you're right. You know, a very funny thing happened. After I made up my list for the party, I invited my neighbors by word of mouth. About a week after Father got rid of our problem, I saw my neighbor ... not the one whose husband died. I caught her outside at her mailbox and invited her and her hubby to the party. Out of nowhere she said, 'Do you believe in ghosts?' I said, 'Why do you ask?' And she said, 'Well, I've got one. It's a long shrill cry that comes and goes.' Can you believe that?"

CASE FILE: AMERICA'S MOST HAUNTED

FOR GOD'S SAKE, GET OUT!
(Pre-release billboard hype for *The Amityville Horror)*

Defending *The Amityville Horror: A True Story* (Prentice-Hall, 1977), author Jay Anson said, "You cannot summon the supernatural on demand as proof." Twenty years after the release of *Amityville,* I investigated a house that rivaled the infamous three-story Dutch Colonial. My nomination for the title of "America's Most Haunted" was a virtual *potpourri of* apparitions, disembodied voices, unexplained fires, bangs, pops, and miscellaneous disappearances. I didn't even try to summon up that lot.

(Transcribed from the tapes of October 15, 1997.) "It's 7:15 p.m. Present are Andrew Nichols, Michael and Zelda Brown; their grown daughters, Nancy and Janet, and granddaughter Terri. Those witnesses to events not in attendance are Zelda's mother, her son and daughter by a previous marriage, and a sister living in Chicago. Also absent are Lynn and Martin, friends of the family." As usual, assumed names were used to conceal the identity of our clients and their friends:

Zelda: "I don't know where to begin. I think almost every day *something* happens."

Andrew: "Just do your best."

Zelda: "Well, we missed a gallon container of milk. We used it the night before and the next morning ... when I went to look for it, it was gone. Then, that night, I was bending over to pick something up off the kitchen floor and right in front of me it appeared again. It scared me so bad I didn't believe it. Does that sound strange? When I was a kid my dad and mother were strict. I wasn't allowed to believe in anything like that, so I didn't believe it."

Andrew: "How long had you lived here before that took place?"

Zelda: "That was 1986, just before we actually moved in. And my granddaughter

Terri...she was five years old then ... she used to cry and tell me there was an old lady in the house aggravating her and taking her toys. One night she said, 'That old lady came and got my baby.' And we never did find her doll; but I still didn't believe in ghosts. She was five then; she's sixteen now. "

Janet: "And then, of course, eight years ago when I was twenty-one I saw something sitting in that chair (referring to the one I was in). It was white, not solid, and had no shape to speak of. .. and it glowed from the inside out. It lasted for about two minutes."

Andrew: "Could you make out who or what it was?"

Janet: "No. It was just a blurry-like light about the size of a small person."

Andrew: "Who all lives here now?"

Zelda: "My husband and I are the only ones. I was sitting in that same chair just the other night and my friend Lynn ... she was sitting there across from me ... and that chair shook so violently. She saw it too."

Andrew: "You told me over the phone that you recently had some work done?"

Zelda: "We had workmen in here to do some remodeling. They couldn't believe it; their tools would disappear just like ours do. And there was no one in the house but them at the time."

Andrew: "Is that when things started to increase ... when the work on the house began?" Zelda: "No. I don't think so. It's just different times. Maybe two or three weeks or a month I don't hear anything, and then all of a sudden it's back."

Andrew: "You heard voices?"

Zelda: "Oh ...all the time. Twice I heard a lady cry. I thought it was my daughter, and she told me no. So I thought she was arguing with her husband, and it was none of my business. So I forgot about it. The next time I heard it there was no one in the house but my husband and me. I couldn't sleep so I thought, *I'm going in the other room so he can rest*. And I heard it and I sat up straight and I thought, *I'm losing my mind, because I'm not really hearing this.* But the crying went on for the better part of an hour."

Andrew: "Where did it come from?"

Zelda: "It seemed to come from the room right over the dining room ... it's a second floor bedroom.

"I can feel it sometimes on this couch. One time, I'd just come across the phone number of this minister I was going to call. My sister was sitting across from me, and it held me down so I couldn't get up. I couldn't move; I was so scared. She thought I was having a nightmare. I said, 'I'm not asleep!' When she came over, it let go. I said, 'I'm going to call that preacher tonight. I'm not going to wait!' But the phone number was gone, and I couldn't remember his name."

Andrew: "Was this someone you knew?"

Zelda: "No. A stranger at my work told me about him ... that if he got permission he could come and probably help me."

Zelda: "Janet and I were sitting on this couch one evening when something sat down with us. You remember that don't you?"
Janet: "It was really eerie. You could feel something sit down next to you."

Andrew: "What has Mr. Brown experienced?"

Michael: "Nothing."

Zelda: "He hears things, but he'll say, 'Well, something fell upstairs,' or there's an excuse for it. He just don't believe it."

Andrew: "You were home when crying was going on, Mr. Brown?"

Michael: "I guess."

Zelda: "The second time he was. The first time he wasn't. I heard it ... he didn't."

Andrew: "Have either of you actually seen anything? I know that Janet and Terri did, but did either of you?"

Zelda: "Well, I thought if I would see something, I'd see something I could identify. But it really isn't. A lot of times I'll just see something coming down the steps. It's just like that smoke (nodding towards the smoke curling off Nancy's cigarette) floating down the steps ..... but denser. One night I looked right there, where that empty chair is sitting in the corner, and I saw an animal. I've never seen anything like it, not even in a picture. It was an animal with short legs and it had sort of like a short head with ... like little horns. It was standing there looking straight at me, its eyes glowing. It was solid. It wasn't like the smoke or anything. It wasn't much higher than this coffee table."

Andrew: "What do you know about the former tenants?"

Zelda: "The lady who lives next door told me that the woman who lived here used to hold seances. They had all kinds of weird people coming in and out all the time. Her children would sit on the porch and watch them come and go. The woman died here. "I wish you could talk to my sister. She's back in Chicago now. Every night she'd hear them talking ... said it sounded like mumbling. She couldn't hear what they were saying but thought they were talking about her. It was just like a movie. There was a man, a woman, and a child."
(I wonder now why I didn't follow up on the mumbling threesome. I was probably suffering from phenomenon overload: sixteen different events and I'd barely begun.)

Andrew: "You told me about your son ... that he slept in the attic and was jolted from his bed?"

Zelda: "It's a bad place. And my other daughter, a child from my first marriage, she slept up there for about a year. She just ... every night she just couldn't sleep. She said she felt it touch her. She'd hear it come out of the closet. She said one night it woke her up ... she was falling asleep with a cigarette in her hand ... and it woke her up. It got her by the foot. It's grabbed my foot several times while I was in my bed."
Andrew: "Have you ever felt threatened?"

Zelda: "One time I did. I was scared to death! Used to, when I'd hear it, I'd call the police."

Andrew: "Did you really have reason to do that, or were you just frightened by the sounds?"

Zelda: "It was the sounds. It seems to be worse when there's snow on the ground ... when there's a lot of snow. I was frightened for myself. I was so scared I got behind the couch; that's where I'd hide. (Turning toward her husband, who had what I suspect was a wider than normal grin on his face) Now this is so ... it really is so!"

Andrew: "What made you so scared?"

Zelda: "Of whatever the noise was. I knew there was no one else in the house but me. The next night my daughter came in, and it was so bad. It sounded like it was on the landing; and we went half-way up the stairs and talked to it. We tried to beg it to just leave us alone."

Andrew: "What did it look like?"

Zelda: "I didn't see it. I just heard it."

Andrew: "What did you hear?"

Zelda: "Just mumbling and groaning and scratching."

Andrew: "But you could tell approximately where it was?"

Zelda: "Yes."

Andrew: "What do you think of all this Mr. Brown?"

Michael: "I don't think there's anything here."

Zelda: "He thinks I'm crazy."

Andrew: "You think it's just imagination?"

Michael: "Yes sir!"

Zelda: "Well, I'll tell you what my mother imagined the other day. It got her dress and pulled it down over her hips."

Michael: "That is when she *fled* the scene" (The corners of his mouth were stretching to reach his ears.)

Zelda: "It really hasn't bothered her until recently. She just didn't believe it either, you know, because she didn't hear it or feel it. But then she felt it pull the covers off her several times, you know."

Andrew: "Your mom used to live with you?"

Zelda: "She did until recently. Then she left. She would never stay with us after that."

Zelda: "My son ... after we moved in his friends wouldn't even come here to visit. Except this one morning ...I guess it was about three o'clock ... his friend Martin came down. He was spending the night in the attic and asked if I'd drive him home. And I said, No.' And he said, 'Well, I guess I'll walk.' I said, 'What happened?' And he said, 'You're not going to believe this but there was no one in that room but me and something got hold of the bed and shook me so hard that I had to hang on to the sides; but I still slid off the bed.' And he would never come back into this house."

Andrew: "How old was this boy?"

Zelda: "Sixteen ... and my son was fifteen when we bought the house."

Zelda: "One time I was in the basement washing clothes and the light clicked off. I couldn't get back upstairs fast enough. Then it clicked back on by itself. And the only way that would click is if you pull the chain. It did that for about two or three days. "And then once I couldn't find some sweaters I'd put in the dryer. I looked everywhere for those sweaters. I couldn't imagine who took them. Two or three days later they were back in the dryer."

Andrew: "You said you had a window that you couldn't keep opened."

Zelda: "That's on the landing between the first and second floor. It would go up ... and it's easy to go up ... and then it would *bang* down."

Andrew: "Does that window have counterweights on it?"

Michael: "Yes." (Only Michael denied ever hearing it slam shut.)

Zelda: "I think another person saw that animal I saw, too. My granddaughter's friend was here one night, and she said, 'Grandma ... she calls me grandma ... where'd that dog go to? Where'd it come from?' I said, 'Honey, we don't have a dog. She said, 'One just ran behind the couch.' She described it and I hadn't even mentioned it to her. So evidently she saw it, too. "And my grandchildren said they saw the bathroom door open and close by itself."

Michael: "Five years ago she had some woman in here to do an exorcism. "

Zelda: "A psychic. She brought some helpers with her (something I was not aware of). I was scared to death when she told us we had seven or eight of them here, because I had just finally decided that ... well, we do have a ghost ... but (twisting to face her husband, while shaking her head) I never ever believed in it."

Andrew: "What do you think is the cause of it?"
Zelda: "I have no idea. I just don't know."

Andrew: "You say there were folks living here previously ... experimenting around with seances and so forth?"

Zelda: "That's what my neighbor told me. I didn't know the people that lived here."

Andrew: "These things occur just any time?"

Zelda: "Right, day or night. If it's really bad, sometimes it starts around three or four in the morning."

Andrew: "Have you ever had a pet here?"

Nancy: "We had a dog; I just loved him. We wanted him to sleep in the basement ... it was so cold outside. So one night he did, and from then on we couldn't get that dog in the basement; we couldn't even get him in the *house.* They tried to drag him in, but he wouldn't go."

Zelda: "He slept out on the back porch with the ice about six inches thick. He got pneumonia or something. We had to have him put to sleep. My son-in-law even tried to carry him down to the basement; he wouldn't go. (Another reluctant pooch!) "I've called the police here so many times I was embarrassed. I just knew it had to be *some body.* I couldn't believe that it could be something that you couldn't see."

Andrew: "The people you had in here, did they say they'd taken care of it?"

Zelda: "They didn't think so. They said to call them if the ghosts weren't gone. One of them, when she first came in, said, 'I don't want anything to do with this house.' She sat on the couch while the others went through the house. There were nine of them altogether."

Andrew: "Was there any change after they left?"

Zelda: "No. It didn't get any worse, but it didn't get any better. When that same group went to the house just two doors from here they had success."

Andrew: "Do you know what kind of things were going on there?"

Zelda: "They were just like me. They didn't believe it. They thought something was wrong with their daughter, so they put her in the hospital for observation. She was about fourteen or fifteen at the time. She kept saying this old lady was coming in there. The doctors couldn't find anything wrong with her. One night she was screaming that the old lady was scratching her stomach. She had scratches all over her stomach; they were pretty severe."

Nancy: "Right after they moved in I was visiting there. I walked in the front door and I heard something moan at the top of the stairs to the second floor. I was kind of reluctant to go back, you know. They'd only lived there a few days, and I didn't want to get carried away ... but it was real to me at the time."

Zelda: "Two or three years ago my mother and I heard some commotion out front. She looked out the door and said,'There's a man on the porch. I looked out at him and I said, 'Mom, that's a man I thought was dead!' It looked so much like him that I couldn't believe it wasn't him. I didn't know what to do, but I wasn't about to open the door... So I thought, *Well, I'll just lock it...* And I noticed he had his key trying to come in the door. And I said, 'Can I help you, sir?' He said, 'No ...I want in!' I said, 'You don't live here.' And he said, 'I do live here. I've got the *key.'* "So I called the police, and I said,

'I don't ... you know, I don't want him put in jail or anything like that but, I said, 'I think he's confused.' When the police came they said, 'What's the matter with you...are you lost?' He said, 'No, I live here ... I've got the key.' The last time I saw him the police were leading him across the street. It really shook me up. He was just identical to the man I knew who was dead."

Andrew: "The man he resembled ... was it someone from this area?"

Zelda: "No. He only lived here for a while."

Andrew: "But you know for a fact that he passed away?"

Zelda: "Oh sure! I know for a *fact* he's passed away. He looked exactly like this person. He dressed like him and wore his cap turned sort of on an angle like *he* did."

Andrew: "And it couldn't have been the same guy?"

Zelda: "It was impossible. It couldn't have been him."

Andrew: "He looked just like a regular person? He talked to you?"

Zelda: "The only thing he said was he wanted in the house. But what really shocked me was when he got out into the middle of the street ..... the policemen had him by the arms ... he jerked away, and he turned around and looked right at me. He just stared right through me. It was unreal!"

Andrew: "How long has this person been dead?"

Zelda: "Five years."

Andrew: "And how long ago did this happen?"

Zelda: "Three years, I think."

Andrew: "And you didn't talk to the police after ... "

Zelda: "No, I didn't talk to them after. The boy across the street said, 'I saw that strange man sitting on your steps.' He said, 'I just wondered if you were going to need any help because he acted like he was sick or drunk.'"

Andrew: "About how old are these houses?"

Michael: "I would guess around the turn of the century."

Zelda: "I know I've talked to a lot of people; there's a lot of homes in this area that are haunted like this, or even worse. In fact, I met a girl in the grocery store and we just got to talking, and I told her that I lived in a haunted house. And she said, 'Oh, you do? We do, too!' She said, 'We're moving ...

we can't stand it.' She said, 'Where do you live?' And I told her and she said, 'Right across the alley from us.'"

Andrew: "Did she tell you what she was experiencing?"

Zelda: "Yeah. She said it disturbed the baby all the time; wouldn't let the baby sleep. They could see a light, like a blue light over the baby's bed ... not a light, but a glow-like."

Andrew: "How about other sounds ... didn't you tell me about doors closing?" Zelda: "Always doors slamming and opening and walking sounds..." Phil: "What about other things you've seen?"

Zelda: "One thing I saw about five years ago. I just came home from work, and Michael was on the recliner, snoring-away. And I thought, Boy, my eyes are playing tricks on me. .. the chair is about this much (indicating a good foot) off the rug ... *and him on it.* I couldn't help it ...I let out a scream and down he came, crashing to the floor." While she was talking, Michael nodded his head. When I asked him what he thought about it, he said, "Yeah, there's been a lot of funny things going on. I guess I was off the floor some (he showed me where the fall had cracked the chair at its wooden base), but I *still* don't believe in no ghosts!"

Andrew: "Aside from what's going on here, do you folks have any special interest in spirits and hauntings?"

Zelda: "I never did. I didn't believe in it. Now I read every book I can get on it."

Andrew: "Can you remember any of the titles you've read?"

Zelda: "Well, of course, Amityville Horror. I think that is a 'horror.' I mean, I just don't believe all that could happen ... except maybe the bees. Just the other day our second floor bedroom was full of bees. I don't know how they got in there, because the windows haven't been opened."
(Actually, flies not bees were found (according to Jay Anson) in an upstairs bedroom in the Amityville house.)

Zelda: "The room at the end of the hallway upstairs is where the bees were. That seems to be the worse ... if there is one worse than the others. When my son came back from Viet Nam I told him to take a nap in there while I went to the grocery store. He wouldn't do it. He said, 'Mom, I'd rather go back to Viet Nam than stay in this house.' "But the worse thing that happened to my son when he was young happened one night in the attic. Somebody tried to choke him. At first he thought his sister had snuck up there and done it, and he was really angry. The next morning his throat was still sore."

Zelda: "Oh, 1 was going to tell you about my sister. She slept in that room ... the bedroom at the end of the hall. And that's the worse room. She said the noises and voices sounded like they'd come out of the vent. It would get in bed with her ... go over the bed then open the closet door. And they were talking all the time."

Andrew: "Was it just the sound of voices going over the bed?"

Zelda: "Yes. Well, she could feel them on the bed. She'd leave the light on all night because she was scared to sleep. Then one night she was reading and she had her elbow resting on the pillow next to her, and it just throwed the pillow over her arm. "It's pulled the covers off my mother and it does it to me. I just pull it back up over me and right away, off it goes. It can go on all night. "But we miss things. It takes things you're going to need. And you might as well forget about looking for them; you won't find them-not right away. Some things you never find. One time it got a can of starch. I was going to do Michael's shirts. We couldn't find that starch, and we looked everywhere for it. All night I heard mom looking in the cabinets and pulling the drawers open. And I said, Mom, forget it. You won't find it.' Then, I guess it was six weeks after that, I heard a loud noise in the dining room. The can of starch was sitting on the table. It's always taking something you need right away. "All of my jewelry disappeared ... all my good jewelry. I even had the police come. They took fingerprints and everything. Two weeks later everything came back. Not one piece was missing."

Andrew: "You told me it knocked the phone receiver out of your hand?"

Zelda: "I picked the phone up to call you and I thought I'd dropped it. Then I picked it up again and it grabbed it...it grabbed my arm way back near the elbow and grabbed the phone out of my hand It hurt my arm ... really damaged it. It seems it doesn't want me to contact you or anyone else."

Andrew: "Well, could I talk you into walking me through the house?"

Zelda: (On the stairs) "This is where my friend from work saw feet."

Andrew: "What did she see?"

Zelda: "She saw *feet* coming down the steps."

Andrew: "No body ... just feet?"

Zelda: "Nothing but feet."

Andrew: "You told me you had two fires go up this staircase?"

Michael: "Yeah. We had one bad fire with ten-thousand dollars worth of damage. No one was home at the time."

Andrew: "Did they ever determine what caused it?"

Michael: "No."

Andrew: "What was the other one?"

Michael: "It was just a small fire."

Zelda: "Up in the attic we have these storage cabinets. Well, a few weeks ago he and I heard the doors on them banging open and shut for about ten minutes. And then later we heard music coming from the attic." (Michael Brown said nothing.)

The barrage was over. I had listened to four dozen chilling and not so chilling incidents. It was a Ghost Detective world indoor record--thus far unchallenged. Only *Amityville* itself–considered a virtual compendium of every conceivable type of psychic event--surpassed it. Included among the manifestations purportedly observed in the "house that hype built" were:

- The presence of an uninvited old lady.
- A couch that moved on its own.
- The sight of a horned creature.
- The feeling of being touched by something unseen.
- Doors and windows that mysteriously opened and slammed shut.
- A dog that refused to stay in the house.
- A levitating family member.
- A really bad bedroom with flies that cluttered up the windowpane.
- Missing valuables.
- Persistent disruption to telephone calls.
- Phantom music.
- All sorts of other uncanny sounds.

If these extraordinary effects seem familiar, they parallel almost exactly phenomena claimed by the Browns, except nobody told them to Get Out!

In my opinion the things reported were far too disjointed and subjective to be taken seriously. Michael Brown had a pretty good handle on this thing when he said, 'I don't think there's anything here.' It's one woman's imagination supported by her family. As for the bees ... they're more likely in their bonnets than in their bedrooms." Then why do I think this case is worth telling? I have included it because of something that took place subsequent to my visit. On Thursday, October 18, 1997, I telephoned the neighbors across the street from the Browns. I was looking for the boy who had, three years previously, witnessed a strange man sitting on Zelda's front steps: a "dead" man the police had taken into custody.

It turned out that the "boy" was now a man living on his own. His grandmother, however, was happy to talk to me about what she'd seen. Her remarks were not tape recorded, but my notes indicate that they went something like this:

Grandma: "I was sitting in the living room and didn't see it clearly myself. .. but my grandson told me what he saw, and we agreed that we'd seen the same. The black and white pulled up in front of our house; then the officers walked across the street. I wondered to myself why they had to park here to go over there. My grandson was outside waiting for a ride when they came back with the man in tow. All of a sudden he gets loose, he spins around and disappears. He up and vanishes! They looked at each other, got in their police car in a hurry and left without him."

The next morning I toyed with the idea of calling the local Police Station to verify grandma's story. I probably should have gone in person, but human observation and memory being what they are (notoriously faulty and unreliable) I wasn't feeling too confident in my source or my mission: I could always chicken-out over the telephone. Eventually, I worked up the courage to call. At first, the officer on duty thought I was a crank. Then, I think to get rid of me he said, "Sorry ... we don't keep that information for more than a couple of years," and hung up.

Looking at it realistically, if what I was told was true no police officer in his right mind would have admitted it. When I told a colleague about it he just laughed: "You didn't give' em your name, did you? Can you imagine what the cops must think of the Browns," he added, "with all the trips they've made to the house over the years looking for phantom intruders? You'd better check your legs my friend. You might find one's a little longer than the other today."

Chapter 3: The Game's Afoot!

"How often have I said to you that when you have eliminated the impossible, whatever remains, however improbable, must be the truth?"
(Sir Arthur Conan Doyle, *The Sign Of The Four, 1889)*

Among the most dreaded experiences we human beings can endure is the ridicule of our peers. And, next to the flying saucer contactee, no one is bombarded by more verbal abuse than the poor soul who claims there's a ghost in his house. No one wants to be the butt of jokes, worst of all those that hint at their mental stability. But instead of commiserating with them, coworkers, friends and neighbors--even relatives--are apt to lampoon the haunted. Psychologists tell us that it's a way for them to cope with their own deeply rooted fears and uncertainties.

Since people will go to almost any extreme to avoid humiliation, it's little wonder that they shun publicity. These are usually steady, unimaginative, unexcitable folks with just one odd experience in their lives. It's easier for them to ignore it than report it. Even when they've reached their wit's end (which is the condition many are in by the time they call me) they're still reluctant to risk almost certain scorn. Only after they're convinced that their secret is safe with me will they begin to talk about it in detail.

Once the idea that they're haunted takes hold, everything seems to reinforce it. Even ordinary things: settling noises, pipes rattling, wind-blown branches against the house, lights reflecting from passing automobiles, the faint sounds of a neighbor's television- things that are always around but rarely noticed are mistaken for ghostly phenomena. Sometimes a series of subtle yet unnerving events does herald the arrival of a ghost. Most of the time, they're nothing more than tricks of the imagination: acts that seem to breed on fear and multiply on attention. Odd but inoffensive incidents should be easy to ignore; and that's exactly what I ask them to do. "When ghosts come a-calling," I tell them, "they leave little doubt about their existence! No speculation is necessary."

Unless they're predisposed to believe in such things, many begin to wonder about their sanity. They invariably start the conversation by saying, "You're gonna think I'm crazy, but..."; or, "I've never been 'funny in the head' and don't believe in it, but..." There's a vast psychological middle ground that lies between our idea of mental normalcy and flat-out madness. Many fail to recognize that you can live through hair-raising experiences without being or becoming deranged. Part of my job is to get that message across.

People ask all kinds of questions. They want a label for their experience. Is it dangerous? Could it become dangerous? Reactions are across the board. At first curiosity, then confusion, annoyance, fear, shock, disbelief, and often denial. Surprisingly, not everyone wants their "visitor" dislodged. Despite the unsettling thought of something roaming about the house, as long as no one is physically harmed or in mortal fear, life with the ghost can settle into a sort of routine apathy. In fact, some handle the business remarkable well. They treat their house geist like a house guest--a member of the family--even going so far as giving it a name.

Is There a Parapsychologist in the House?

Those who have the desire and the courage to do so reach-out for help. Except for people like me, not much is available. Primitive man, I'm told, didn't fear ghosts because he didn't fear, or perhaps, understand death. Death was a time for celebration. The newly departed were on an adventure to eternity, not a trip to oblivion. In the era before organized religion, magicians and sorcerers were given the task of explaining and expelling bothersome spirits. They were seldom sent off to eternal damnation, just told to go bother somebody else.

Modern man is haunted by the specter of death and spends an enormous amount of effort, conscious and unconscious, denying his personal role in it. Somewhere along the way he's lost belief in religious survival. Still, when he finds himself faced with "other worldly forces" he's likely to seek out twentieth-century's counterpart to the sorcerer: the clergyman. If one form of orthodoxy, the Church, fails him, which I'm sorry to say is often the case, he may turn to another: the medical practitioner. When both disciplines fall short and he's left thoroughly befuddled, he may turn to people like me.

The public lacks knowledge about the subject. It's an ignorance that extends to the clergy and the medical profession: a painful revelation for those beleaguered folks who look to their ministers and doctors for help. Most are uncomfortable, or just plain hardheaded about dealing with psychic matters and give the topic short-shrift.

There are clergymen who, for all their professed conviction in the existence of the spirit, neither believe in, nor will they even tolerate talk of non-ecclesiastical visions. It may come as a surprise to some of my readers, but there's no scriptural basis whatsoever for belief in an "essence" that consciously survives death.

Victims fare no better with their physicians. Some treat the haunted as if they were on drugs. Some are called liars. Others pre judged as mental misfits. Because of the vague and confusing nature of the subject, few medical practitioners are willing to listen while their patients ramble on (incoherently, according to them) about entities and evil spirits. Battling the ravages of disease is an all consuming job. Little time or tolerance is left for "cockamamie" ghost stories. But discrediting their testimony, or passing them on to a shrink" without learning more about the event is not only premature, it's unprofessional. Telling a person that their experience was a figment of the imagination is a non-explanation and may cause more harm than good.

It could be argued, I suppose, that these encounters lie outside the domain of religion and medicine. At the same time, there's a crucial need for help to interpret what's happening to people. After years of being asked the same questions you'd think these experts on the human condition would be prepared to make some contribution to an understanding of the problem. Grasping for help, the haunted call the police, fire department, newspapers, radio and television stations, social service agencies, universities. A few call 911 (which, of course, is a no-no). Some confide in friends and neighbors; a good many keep their embarrassing problems to themselves.

It may be difficult to locate, but aid does exist. In addition to me, there are groups across the U.S. that are able, or profess to be able to give some degree of help. They range in expertise from The American Institute of Parapsychology in Gainesville, Florida (of which I am Director), and The American Society for Psychical Research in New York City, to groups that study metaphysics, Mind-Science, Human Dimensions, and so on.

A small number of research societies have trained investigators who make field visits, although their abilities vary greatly. Many, unfortunately, turn out to be no more than chroniclers of ghost stories. Their reviews, if they even bother to make them, wind-up gathering dust on a shelf somewhere. Due to the lack of adequate funding only a few universities have departments of parapsychology. As of this writing, only a couple hundred parapsychologists practice their profession. Most are doing laboratory experiments with ESP and PK. A handful of them and a few more qualified investigators, spread thinly across the U.S. and Canada, actively investigate ghosts.

Since I seldom accept one absent of corroborated testimony, my cases have consisted of households in which there were two or more residents. Women between the ages of twenty-five and fifty-five headed approximately two-thirds of these homes. Some were separated from their spouses; some widowed. Many were divorced. (As mentioned earlier, women without a man in the house seem to be more prone to experiencing phenomena.) The rest: married couples; single women living together; and single men and women who, as is the growing custom, lived together without the benefit of matrimony.

The majority of my clients were slightly above average in education and earning power. Better than two-thirds had children living at home; a normal, if somewhat reduced social life; and, as far as I could tell, were well balanced mentally and emotionally before their experience. (Not that they subsequently went to pieces; but, for obvious reasons, some were not at their emotional best in the beginning.)

Most followed no particular religion. Among those who did, however, were Protestants (a few Fundamentalists included), Catholics, Jews, and so far, one Moslem. Naturally, the greatest numbers of those with religious affiliations were Christians, since Christianity represents the majority faith in this country. Many had only a passing interest in their faith and, if they went at all, were probably below average in church attendance. Few had knowledge or interest in the Occult or in things supernatural before their bizarre adventure.

Except for the large percentage of women who head households (not a surprising statistic when you remember that the divorce rate has climbed to around 50 percent over the past twenty-five years), my clients represent a typical cross-section of the population. The numbers show, I believe, that anyone is fair game for an out-of-this-world experience regardless of their social, economic, or educational status. There does appear to be some relationship between an experient's level of sophistication and the account's level of *strangeness.* Those aware of the ways of the world are less apt to tell their really far-out encounters than the less sophisticated. Naturally, this leads to a distortion in the statistics of "very strange" cases compared to "mildly strange" ones. Those observers better educated and more eloquent may be reserved in their choice of words to describe what they've seen. Yet in all my adventures and in all the case histories I've read, reports that range from mild infestation to wild manifestations are distributed fairly evenly across a broad demographic spectrum of the population. I can make that statement with some confidence because of the large number of phone calls and the bags full of mail I've received over the years.

They come from every walk of life. The majority are honest and well meaning, and, as a rule, don't concoct tall tales. Until their everyday lives were turned upside down they acquired their knowledge of ghosts from movies, television, novels, and the press. Essentially, their attitude mirrored the communication industry's treatment of the subject. You might assume then that out of the ordinary experiences would reflect the things they'd seen, heard, and read in the media. Not so! In case after case, they relate details that correlate with real phenomena--phenomena, which in the main have been revealed to the public only in the relatively obscure books of equally obscure researchers like me.

As you may have guessed, every time a new novel, movie, or TV miniseries comes out extolling themes of demonic possession, or sexual harassment by a Poltergeist out-of-control–my phone rings-off the hook. The latest round followed the inauspicious exorcism of a pathetic teenager on ABC's 20/20.

Few of my callers are experiencing what psychologists term "event-level reality" (the real thing), but I'm glad they call. It gives me a chance to allay their fears about being chased by the Devil, being gang-raped by ghouls, or seeing their kids disappear into the television set. It also gives me the opportunity to urge them to get counseling, if they haven't already gone that route. The things they tell me makes it easy to spot the imagined fears of the highly impressionable, and the delusions of the unhinged.

Separating the Candidates from the Kooks

Corroborated testimony, or the lack of it, is the number one factor in my decision to take or pass on a case. Having others bear witness is no small matter; I seldom accept one without it. It's not that a sole witness is incapable of observing genuine phenomena. The difficulty arises from the fact that purely subjective experiences may be a sign of mental illness. Without confirmation or solid evidence, I may find myself listening to the delusions of a troubled mind. Therefore, before offering my services, I must be certain that a genuine need for them exists. I may not require that the prospective client prove his or her claims before I'll show up; but I do insist on corroboration of the telltale signs of a haunting in progress--not just confirmation of the overwhelming anxiety they're bound to give rise to.

It's difficult to consider intangibles like anxiety, apprehension, and the idea that someone's watching you as reasons to take a case. These fears are important only when part of a scenario in which substantial physical effects have been confirmed by others. Emotional outbursts, brought-on by deep feelings of anxiety and dread, are often contagious--that is, easily picked-up by others. This false "cycle of fear" is frequently started by an authority figure: a parent, older sibling, or strong-willed person. A way to remove myself from the situation without appearing callous to the needs of the family is to insist on corroboration of the physical actions of haunting entities; not merely proof of the strong emotions they evoke.

It may be true that you can tell when a pathological liar is telling the truth by watching his lips--they don't move. But you can generally tell when people are lying: They're too brief; they rub their hands together, or scratch themselves; their pupils shrink; they clench their jaws, cross their arms in front of them, or lower their heads. And, of course, the old standby--they don't look you straight in the eye.

When you're a self-proclaimed authority on anything of public interest you set yourself up for all kinds of experiences with all kinds of characters. While back-peddling along the foul-line, a veteran third-baseman once said, "Self-defense is the better part of valor." Backing-off a little, putting some distance between you and the contactee at the outset, is only prudent. Screening prospective clients before agreeing to take a case is one method of defending against the "line drives." Screening helps cut down the chance of my winding-up in some 'nutcake's' house.

Again, the majority of those who would make use of my services are serious-minded, level-headed individuals; a few are not. The first task is to evaluate reports before investigating them personally. I use the telephone to gather as much ground-work material as possible about both the event and the person or persons affected by it. It's not an airtight method. Often, those who have difficulty lying to your face have no trouble fibbing over the phone. But what I learn is usually indication enough of whether an investigation would be helpful, or a potential waste of time.

It's a simple fact. Once they're over the "you're going to think I'm crazy" stage, those in need of help have no difficulty in telling their stories. They have an overwhelming urge to do it; to be assured that they're not alone; that others have had similar experiences. Anyone willing to lend a sympathetic ear, guarantee confidentiality, and offer something constructive will hear their tale. Interviews range from a simple exchange of ideas to a full-fledged study. My callers want to know what they're up against. Most, but not all, want it to go away: the sooner the better. A few reports are of disturbances already ended. Naturally, I can only take cases in progress, but I do listen and try to help them understand what I think they experienced at the time.

Others tell of active hauntings in the home of a friend or a relative: "I wish they'd call you," they say, "but they don't know how *haunted* they are." I never make unsolicited calls or visits. There's not much I can do to help them unless these people request it themselves. I ask for basic information to screen out the waste-of-timers and determine if an in the field investigation is worth pursuing. I need to know:

1) The details of the phenomena and the names of those who witnessed them.
2) As much information about the house and its location as possible.
3) If there is a known history of psychic abilities among the residents.
4) If any of the witnesses has a history of mental or emotional illness.
5) If they are taking medications or drugs, or if they're alcoholics.
6) What they think is facing them.
7) Why they believe the incident is supernatural.
8) What their religious affiliations are, if any.
9) What preconceived ideas they have about such things as ghosts and hauntings.
10) What they expect from us.

Before taking on some of the more complex ones, I may ask the caller to write-up their experiences and keep an extemporaneous register of observations. Not everyone is willing or capable of maintaining such a log; the less earnest won't even try.

Of the dozens of phone calls and letters I get each year, only a few results in a field investigation. Most contactees are experiencing low-level Infestation, effects that peter out on their own if given enough time. Others, although equally innocuous, will remain in a non-accelerated holding pattern for what may seem like an eternity. My presence is seldom warranted in either case.

One point I want to stress involves the advice I give those who have low intensity, annoying phenomena. I tell them to be patient; to wait their haunting out and, if possible, not focus on it. To each of us our home is sanctuary. Unless you've lived with them, it is impossible to understand how nerve-racking these effects can be. Ignoring them, even the subtle variety, isn't easy. It's all the more difficult when you're frightened for the safety of your family.

Yet I know that if my clients spend time concentrating on the problem, a type of subliminal ammunition is created that reinforces mind-projections. The degree of the haunting (it's intensity and length of duration) is easily influenced by what they say and how they act. Beyond that, those who become emotionally distraught seem to be far more vulnerable than those who manage to stay calm. Whatever its cause, focusing attention on what's happening: wringing your hands, ranting and raving, merely discussing what they may or may not be, tends to enhance and accelerate them. The thing "grows" with each telling, probably because human beings tend to embellish facts a little, and these fictitious additions somehow program the mind to greater feats. The patient wait-and-see method pays off when contactees report an end to the disturbance; a slow-down in its rapidly expanding intensity; or, at the very least, a significant lessening of its effects.

Psychic Triage

More and more, I find myself performing "psychic triage" over the telephone. I listen closely for clues to their emotional condition. If they've managed to adjust to the fear and the chaos, there's no immediate need for me. On the other hand, if they're beginning to show signs of panic, coupled with remarks like, "I just can't spend another night in this house," I make an on the scene appearance pronto.

Not that there's a great deal I can do for them on the spot. I can, however, offer a quick, if not permanent "fix" by pointing out that of all the people in the world they have not been chosen for persecution by the "Mysterious Unknown." Even if they have, history is significantly absent of any harm befalling victims of haunted houses. Speaking to them in person does seem to alleviate their fears, while it creates a useful relationship with the family if I have to come back.

Chapter 4: So You Wanna Be a Ghost Detective?

Of the study of ghosts and hauntings psychologist Carl Jung said:

> *"Spirits do exist, they are inexplicable, genuine manifestations of the unconscious mind and like dreams, are a first class source of information about the viewer's mind .... What we are dealing with is a vast and half-lit area, where nothing seems believable but everything is possible. Therefore, before it is possible to arrive at a half-way solid judgment, one must have observed a good deal personally, heard and read many stories and tested and questioned witnesses." (Source: The Psychology of The Occult and The Psychological Foundation of Belief In Spirits. Quoted in Memories, Dreams, Reflections, Random House, Inc., 1961.)*

For over a quarter of a century I have studied haunted houses and the people who inhabit them. I've read all the material I could find; interviewed hundreds of persons and listened sympathetically to their myriad tales of woe. And, even though I'm a parapsychologist, not a physician, I have never failed to abide by the most important rule of the Hippocratic oath: "*Do your client no harm!*"

As a discipline, parapsychology incorporates the study of philosophy, theology, abnormal psychology, anthropology, chemistry, biology, and physics. Yet any mature student of human behavior armed with a tape recorder and a whole lot of patience can research ghosts. Advanced education is not a prerequisite; holding a Ph. D. in the behavioral or physical sciences doesn't bestow any special skills nor assure success.

In addition to patience and an ability to judge people, he or she needs a working knowledge of psychical research: the journals, case histories and biographies of the individual researchers. And most of all, active contact with those who have experienced spontaneous ghostly outbreaks. The investigator must be an authority on the phenomena. He must understand the patterns and be able to compare each incident with the classic cases of the past. He must have the skills of a police detective: the ability to spot inconsistencies; judge the credibility of observers; recognize evidence of fraud. He must know how to look for normal explanations for seemingly uncommon events and to spot signs of mental distress.

Most parapsychologists are trained psychologists. All good researchers are well grounded in the subject, since mind-related phenomena are almost always responsible for what happens. At the same time, we need to guard against jumping to psychological conclusions. Not every experience is a sign of credulity, illness, or deception as some mind-curists would have us believe.

The investigator must have a depth of experience to fall back on when facing complex cases, for the majority will be complex. Without a sound knowledge of the literature and on-the-job training he'll have little to offer. Nothing could be more embarrassing to the fledgling "ghost-detective" and unsettling to his client than for him to be just one of the spectators at a haunting.

Psychic Counseling

Many who come to me have fallen through the cracks in the psychological services system. Therefore, some understanding of counseling is important. I have said that I operate like a therapeutic folklorist. Therapy begins the moment I take what they tell me at face value. I don't accept it blindly, but I do listen. And once I'm satisfied that they're telling me the truth as they see it–I continue to listen.

Advice is my most important service, particularly in cases of Random RSPK where the effects end quickly, while the memory lingers on. Counselors, often clinical psychologists, work with those who are unable to adjust to their "ghost," or to the consequences of their psychic abilities. I suspect that a number of my clients have the potential for ESP and PK. A few may be latent mediums as evidenced by the spontaneous phenomena that occur around them. Although I'm adept at identifying with their doubts and fears and helping them through their crises, I usually refer them to others to administer treatment.

A potentially serious condition facing some is what psychologists call "post-traumatic stress disorder" (PTSD). Just as those who suffer through warfare combat or through other life-threatening situations: accidents, illness, being held hostage, etc., people with nasty hauntings often display symptoms of excessive stress. They, too, often need more help than I can give them.

Courses in parapsychological counseling are offered at the American Institute of Parapsychology. Many universities in this country and in Canada have crisis counseling and intervention programs--programs useful in helping victims. Some parapsychologists have recognized this need as well, calling our efforts "clinical parapsychology."

Ghost Hunting Gadgetry

In the wake of *Ghostbusters* and devices like "zapper guns," my tools seem tame: audio cassette recorder, video camcorder, digital and 35 mm cameras, flashlights, magnetometer (EMF Meter), digital thermometer, ion detector. They're all so unimaginative. I've said that I rely on the testimony of witnesses. Sometimes, however, if a case is particularly interesting, I may administer psychological tests.

If the case is a real "hot one," I may bring out a bag of gimmicks that includes strain-gauge plates to detect footsteps; ultrasonic, microwave, and infrared detectors: state of the art burglar detection devices; even seismographs sensitive enough to hear the "Abominable Snowman" lumbering along half-way up the Himalayas.

Most amateur ghost detectives use ordinary means: tricks of the trade devised a hundred-years ago by members of the Society for Psychical Research. Sciences' original ghost hunters would catch fraudulent mediums with the most unsophisticated tools imaginable: a roll of adhesive tape and a spool of thread stretched across the entrance to the seance room to keep accomplices out; a bag of flour, the contents of which were scattered on the floor. Anyone entering or walking across the darkened room would leave telltale footprints: as my correspondent's father, the shotgun wielding ghost chaser, hoped to prove. The adhesive tape had another use. So skilled were some mediums--and so intent were researchers to eliminate fraud-it was often used to cover bodily orifices. In that way no hidden items called "Apports" could be retrieved from unseemly places.

Precautions against fraud are in order when the honesty of a medium or sensitive is in doubt. Anyone suspected of bilking the public, whether they charge or not, should be checked out with the best equipment available; there's a legitimate place for it in the laboratory, as well. Also, many universities sponsor scientific groups that are obliged to come up with "proof" merely to validate their own efforts. Extensive use of electronic surveillance in private homes, however, may be overdoing it. Researchers who use devices to get verification of a haunting argue that they're trying to catch the ghost by registering and measuring its physical properties. Spiritualists affirm that ghosts have no physical properties; that they're not of the physical world. Perhaps not. But they do affect the environment in physical ways. If the phenomena they create can be quantified, then I'm for it; but not at the expense of my clients.

I have to be on guard for fraudulent acts, but my primary job is not to prove the existence of phenomena; I couldn't even if I wanted to. The truth is there are many more self-deluded, over imaginative people out there than there are conscious frauds. As a result, I spend far more time filling in as a psychiatric social worker than I do as a ghost cop. Haunted houses are chaotic places. I'd rather give up the hunt than employ an array of equipment that might upset what fragile psychological balance remains in them.

When the Ghost Detective Comes

Interviewing prospective clients over the phone is easy. Doing it in person is a genuine skill. One method separates each witness from the influence of the others, with often surprising results. During group discussions the more submissive; a meek spouse, children, live-in grandparents, etc., will tend to support what the stronger-willed tell me. The subservient, or those who hope to avoid confrontation may unreservedly agree to whatever is necessary merely to keep the peace. Their testimony often falls apart when I question them alone.

I don't think it's conscious dishonesty. Memory can be tricky. It can make us think we've taken part in an adventure that, in truth, we shared through the experience of others. By themselves, many confess *they* didn't observe anything; that they merely heard about it, or were nearby when it happened. Obliged to placate the more dominant family member, or reluctant to embarrass a loved one in front of strangers, they feel compelled to confirm it. To be of any value, direct and corroborated information must be consistently accurate.

It's not surprising to find a few discrepancies when comparing what they tell me. Not everyone can recall events in perfect detail. As long as they apply to unimportant details I have no reason to be suspicious. An old trick of interrogation involves giving them the once-over twice, i.e., asking the same questions over again. The idea is to rephrase and disguise them slightly to see if they'll change their stories to any degree. People who have experienced genuine phenomena rarely do. Doubling-up on questions not only helps to verify the responses, it also tends to jog the witnesses' memory.

I know that crucial parts of the experience will be embellished. It happens during eyewitness accounts of emotion provoking incidents. For added emphasis they may exaggerate what they've seen and heard. By the time we've reached the second phase of questioning they've usually calmed down enough to give an accurate account. All this third-degree grilling may sound a little like Gestapo tactics, but it's necessary if I'm to get the straight scoop.

I ask each observer for a chronology of events. I need to know if all those present observed the phenomenon; if there was a pronounced change in room temperature; if anyone sensed physical, mental, or emotional impressions, or felt threatened. It's important to learn if each observer was fully awake at the time. I ask if the haunting happens only at a certain time, in a particular room or area of the house. I need to find out if anyone has had similar experiences in the past. And where apparitional or ghostly forms have been seen or heard, if an observer could positively identify them--or at least tell me if they were human in form.

When circumstances require it I draw a floor-plan layout of the house. It allows me to pinpoint each individual's location at the time of the disturbance. It reveals the person, or persons unaccounted for (important in cases of physical phenomena). Not that such an absence in itself proves anything. It's only prudent to guard against fraud, particularly when children are involved. Youngsters often have the perfect motive (an attempt to gain attention) and ample opportunity to knock on walls or throw things when no one's looking.

I attempt to discover if the location has a known history of tragic or peculiar events. A neighbor is sometimes my best source of information. The fact that a house has a reputation for being haunted will not endear it to most home buyers. As a result, the current owner may be unaware of its off-beat past. A few years ago, a researcher for the television program *Inside America's Courts* called me for my input on that point. The popular show was about to do a piece on the new real estate disclosure laws proposed by several states, mainly to stop this practice. Where they become law, owners will not only be required to disclose that they've got termites, they'll also have to confess to ghosts around the house.

Occasionally, I come across a house that's already been investigated. If the psychic sleuth was not up to snuff the residents may complain that he or she left the place in worse shape than it was before they arrived. It happens when well-intentioned amateurs use the "elephant gun" approach: applying the tools of the seance and exorcism to what was initially a simple case of mistaken identity. Somehow their tinkering has increased its intensity, making the job of cleaning up after them more difficult.

An Early Warning System

Animals react to ghosts in much the same way as they do to live intruders and impending earthquakes Their acute sense of hearing, or a reaction to changes in electrically-charged particles may be working, and not necessarily a sixth sense. Unlike Homo Sapiens, who might be picking-up subtle impressions from each other, the oddball actions of a pet suggest it is reacting to something real. They may be more sensitive to energy patterns and haunting phenomena than their masters because they don't repress them from consciousness as we tend to do. Whatever the reason, they're likely to notice psychic forces long before we do.

Michael Fox (the one who holds a veterinary degree and a doctorate in psychology, not the time-traveling actor) thinks some pets, principally dogs, are, "Very human--sometimes uncannily so." In his book, *Understanding Your Dog* (Coward, McCann & Geoghegan, Inc., New York) he hints at canine super sensitivity. Although Dr. Fox is skeptical about the subject of animal ESP, after hearing some of the stories about pets--especially the pooches-and their reactions to "ghosts," the theory seems plausible.

Some Natural Explanations

During my tour of the house and its surroundings I'm careful to look for things that might provide a logical explanation for the disturbance: the settling of the building; sewer gas; reflections of sun, moon, or passing headlights; and so forth. Once a patrolling coast guard cutter's light was mistaken for a flying saucer skimming along Tampa Bay. At night, as the boat passed by, its searchlight would sweep back and forth from shore to shore. From my client's vantage point on the bank, the oval pattern of the beam falling on the river did look suspiciously like a low-flying disc. Mechanical malfunctions of plumbing, electrical or heating and cooling systems, as well as acts of nature: lightning, ice cracking on the roof, minor earth shifts, underground streams and so on, may also play a role in haunted houses. Sometimes nature's critters: birds, squirrels, raccoons, or mice are responsible for unusual sounds--especially in walls and attics.

A howling breeze can create all sorts of scary noises. The literature of haunted houses is replete with stories of "voices on the wind." In the early eighties, I investigated a house that was being entertained by violin music. When I got there he learned that it wasn't exactly music, it was more on the order of two or three notes. Finding nothing to account for the sounds inside the house, I searched outside and spotted the source at once. In the backyard, threaded through a fork in an old tree, was a telephone line. In time it had cut a slot in the bottom of the fork. When the wind came up the tree branch would move to and fro, and the phone line would slide in the slot. The tree had become the "bow," the telephone wire the "string," and the house the "sounding box"; the result -violin music.

I always check for prescription and over-the-counter medications and discretely inquire about problems of abuse. It should come as no surprise to the reader that in a culture dominated by drugs there are scores of reports attributable to their misuse. Their reading material is of interest. Works on metaphysical subjects, on psychical research and the Occult, may be a sign of a more than casual knowledge of the subject. Someone whose imagination is easily aroused, or who's trying to mislead me could be getting their material from books.

There are certain psychological and physiological effects that may account for hauntings. Earlier, I mentioned hallucinations: false perceptions with no basis in reality, and illusions--mistaken or distorted ideas of the external world. To these add delusions: false beliefs that influence otherwise normal people. Even the levelheaded can meet with an occasional delusion. It's only when they become persistent that they indicate mental illness.

One Saturday morning, a perplexed little old lady and her son Lenny gave me a neighborly call. She could barely contain herself after taking all she was going to take from an "ornery spirit."

"I don't know what to do anymore. I bought this nice pair of shoes ..... put them in the closet and the next morning they were all used up. The bottoms were scuffed and the heels worn down. I don't know whether to buy another pair or not. What if it keeps on doing it?"

"Are you certain they were in that condition the next day?" I asked. "I mean, could it have been sometime later?"

"It could've been ... but I don't *think* so."

I'd never heard of a ghost who was hard on shoes. Most of them sort of glide along, don't they? Because of her advanced age, I turned to Lenny, hoping he might be going along with the story to humor her. But Lenny was no help. No youngster himself (probably in his fifties), he just nodded his head up and down.

Hypnagogic hallucinations (visual images of persons and things seen just before we doze-off), although bizarre are not abnormal. A few years ago, an utterly distraught middle-aged woman complained, "Rows of dead relatives are visiting me just before I fall asleep at night." Terrified they'd come for her, she appealed for help. Before the strange parade began she had gone through a period of extreme anxiety over her own mortality. Apparently, macabre thoughts had revived memories of relatives long-since deceased; just before sleep she'd see them as they appeared in old photographs: "I'm awake, I'm not dreaming," she told me. "They don't say anything. They just float by in their picture frames-row after row ... then take their placc, onc by one, on the wall." Bedtime had turned it into a panoramic screen for her mind to project its fears and fantasies upon. I explained about the psyche's ability to create vivid hallucinations. I recommended she take a more active roll in life: go back to church, join a social club, spend more time with family members who were still alive. She needed to stop dwelling on death; to use denial as a shield from life's realities, which, of course, is what most of us do every day to keep our sanity.

Hypnagogic hallucinations are experienced at the start of sleep. At its terminus are Hypnopompic visions, the half-hazy images seen just before awakening. These "phantoms" arouse people in the middle of the night, or else are there to greet them first thing in morning. They're not dreams, not technically anyway. Like their Hypnagogic counterparts, "wake-up" materializations can be all too corporeal to their drowsy-eyed observers. For the past forty years or so tales of elves and fairies have been replaced by encounters with ETs. Recently the "wee people" were seen and heard in these parts. A fifty-year-old man, whom I'll call Bob Kelly, said there was a whole gang of 'em in his house. "They were making one hell of a racket up in my bedroom," he said calmly. "I was downstairs on the divan watching television when I heard it...sounded like a bunch of squirrels having a party. I ran up there, and as soon as I got to the top of the staircase ..... they stopped. Everything just stopped!" As he headed down again, he heard the sound of high-pitched giggling behind him. "I spun around and barely caught sight of these little people running under my bed. I lifted the edge of the bedspread ... but they'd vanished."

Several days passed without incident, and my caller was beginning to wonder if he'd dreamed the whole thing. He was lying on the divan again one morning (the day he called me), "resting my eyes," as he put it, when noises inside the fireplace woke him. "I thought some poor critter, like a bird, was trapped," he said. He rolled over to face the opening. And there, hanging upside-down with his head sticking out, was a rosy-cheeked little man, the tip of his green, pointed hat nearly touching the hearth below. With his arm outstretched, he was gesturing toward Bob, "laughing his ass off at me." Was Kelly pulling my leg? It's possible. What's more likely is that he was "resting his eyes" prior to both encounters. His diminutive visitors were probably hallucinations; the consequence of Hypnopompic illusions.

As mentioned previously, sometimes people experience negative hallucinations called Inhibition. Inhibition tends to block out anything in our visual field not being concentrated on, and to tune out noises--chiefly monotonous ones. Omissions of sensory input can easily lead to a misinterpretation of a natural event. The Ideomotor effect is a conditioned response caused by an intense anticipation. Some people have an overwhelming need to believe in the supernatural; this alone may be enough to stimulate ghostly activity, i.e., projections of mind-constructed phenomena. Automatic responses can result from the words we use, and even from our thoughts. Both are, in part, conditioned stimuli that can affect us much the same as a bell elicited saliva automatically in Pavlov's dog. The more people insist that their house is haunted and the more fuel others add to that assumption, the more apt they are to attribute all unexplained events to their "ghost."

Physiological causes for the uncanny include: "phosphenes," those sparkling spirals that appear after we rub or get punched in the eye; "autokinetic effects" where objects at a distance appear to move due to the constant movement of our eyes; and biological changes in the cells called "afterimages." After staring at an object for a while (a light source, in particular), its image remains when we look away. The bright semicircle of light that fades away gradually after a camera flash goes off is an example. Although it may appear otherwise, physiological effects in and of themselves never prompt an investigation. Things like phosphenes, autokinetic movements, and afterimages are seldom present in genuine hauntings. Yet, together with psychological hang-ups, fraud, and mistaken identity, they're often cited by orthodox science as the basis for all such reports.

For reasons I do not fully understand psychic disturbances sometimes end, not in response to anything I've done, but simply because I show up. Acting as a sounding board, calming my client's fears, and taking the time to explain what I think is behind the haunting seems to get rid of it. Cases like these must be pure exhibitions of the mind. Mental health field specialists know that psychological problems are much more apt to resolve themselves when clinicians make house calls than when patients are brought into a center for treatment. Apparently, my arrival on the scene acts as a catharsis. It helps to purge their hangups and bring down the curtain on the psychic show.

It's only human, I suppose, but not everyone is happy with a natural explanation for their weird encounters. Some reject it outright, even when a non-ectoplasmic solution should have been patently obvious to them. Many prefer to hang-on to the supernatural, to the idea that an incorporeal being has invaded their privacy. Others, especially the more devoutly religious, vehemently disagree with any theory that excludes the demonic as a finding. According to Donna Larcen and Colin McEnroe ("Hooked On Mysteries," *The Hartford Courant*, 1992): "The solution to any mystery has two parts. The first is a set of facts that constitutes a proof. The second is a willingness to believe those facts. The willingness is often harder to get than the facts."

Separating the Fact from the Folklore

Some ghost stories--because they're borrowed--can never be checked-out. The folks who tell them are spellbinders. From time to time their "marvelous strange" stories make the rounds. In one, a lone male driver (sometimes a cabby) picks up a female hitchhiker late at night. The young woman thumbs the ride near a bridge or crossroads. After taking her home, the driver finds some reason to go to the door (in the case of the cabby, to collect his fare). He rings the doorbell and asks the gentlemen who answers, "What happened to the young lady?" Whereupon he learns, "She was killed years ago at the very spot you picked her up." What's more, "You're the umpteenth person who's tried to bring her home since the tragedy." Folklorists call these cock-and-bull stories, "foaftales" (friend-of-a-friend tales). They're nearly always borrowed adventures experienced, not by the teller, but by someone who related it to them --no doubt to whom it was related, and on and on.

There's another popular foaftale concerning three young men who paid a companion thirty dollars to spend a night alone in New Orleans' St. Louis Cemetery #1. All he had to do to earn the money was stay in the tomb of Voodoo Queen Marie Laveau. His new friends told him to bring a hammer and nail with him. Why? To prove that he'd been there he had to drive the nail into her final resting place. Just after dark they accompanied the intrepid fellow to the cemetery. Armed with hammer and nail, he scaled the wall and disappeared into the blackness, headed for the infamous Voodoo Queen's tomb.

At dawn, the trio was still waiting for him. When the caretaker arrived and opened the gate they rushed to the vault. And there he was, lying dead at its base. He'd made it in all right and hammered the nail as instructed; but something had gone wrong! We can only guess what followed. By the look on his lifeless face, he'd been seized by overwhelming terror. Trying to escape the tomb, he must have thought someone had grabbed him from behind and wouldn't let go. When they found the courage to examine his body, the instigators of this morbid prank saw that he'd driven the nail not only through the crypt, but through his coat as well! It's a tale told often. The last time I saw it in print was in Richard Winer and Nancy Osborn's, *Haunted Houses* (Bantam Books, Inc., New York). According to the authors, they heard several variations on the same theme while researching material for the book. Supposedly, the incident took place in the belfry of the Aquia Church in Stafford, Virginia, as well as in a haunted house on the campus of a Spokane, Washington college. Recently, I was told it happened near the campus of Stetson University in Deland, Florida. I've heard the "dead hitchhiker" and the "nail in the night" stories before. Listening to these secondhand tales is part of the price of doing business.

Humanistic Parapsychology and the Person-Centered Approach

Like a therapeutic folklorist I usually am not investigating a case to prove or disprove anything. Consequently I could be fooled, especially in the opening phase of an investigation. Charles Lamb, English essayist, said, "Credulity is the child's strength, but the adult's weakness." It is unlikely that my adventures have been totally absent of client dishonesty. I have no foolproof way of guarding against the pathological liars and psycho-ceramics out there. Even so, no one spends years studying these things unless he's absolutely convinced that there exist authentic effects caused by other than normal and natural means. I do not suffer liars and fools easily; credulity is not my heel of Achilles.

I employ an "experience-centered" approach which I refer to as “Humanistic Parapsychology.” I base my conclusions on the testimony given by the residents of the troubled house and judge it critically on its merits. Does it fit the patterns I've seen before or those reported by other investigators? Is there an equally acceptable, natural explanation for the things that only seem to be inexplicable? Does it completely defy logic?

I am rarely privileged to observe a disturbance in progress. Haunting entities and Directed RSPK (PK with ghosts) are notoriously shy about command performances. Psychic Imprints and Random RSPK (especially the Poltergeist) are a little less evasive, but not much. Even in the absence of first-hand knowledge, once I've passed the preliminary stage of my visit there's little reason to doubt my client's honesty. Why? Because of their unquestionable sincerity, the lack of a motive for fabricating tales, and the fact that more than one person corroborates what they tell me. Moreover, as pointed out earlier, many complaints are of unique incidents not included in the popular mythology of the subject and, therefore, unknown to the public.

For myself, I'm satisfied with *Whately's Law of Evidence* which states: "If enough independent witnesses agree upon the characteristics of an observation when we have ruled-out the possibility of collusion between them, then the observation has a high likelihood of being genuine and not falsified"

Whatever ghosts are, whether they're mind projections; spirit materializations; or shadows of things that defy explanation, the things they do--the tumult and turmoil they cause--are real. It's impossible for a careful observer to believe otherwise. For once I'm satisfied that a natural interpretation, mental aberration, or out-right fraud can't account for the things observed--no other reasonable explanation is left.

How Soon They Forget

Before continuing, I want to touch on what is to me another amazing aspect of Spectrology. It is the effect which in time causes observers of phenomena to disavow their own testimony. Since I believe my patrons are both the culprits *and* the victims of these melodramas; that psychic disturbances are the result of subconscious projections, my client's outlook: the way they react to their "ghost"; the content of their dreams; family interrelationship, etc.--their overall mental and emotional well-being--is an important consideration. Because it continues to be important long after I've "cleared" the house, I manage to override my curiosity about what's become of them: I don't get back to them for many months after our visit.

I am inextricably linked to the haunting. After it's over, clients begin to think of me as an integral component of it. Disturbances serious enough to require my presence are often serious enough to leave them with a plethora of bad memories, and worse, the fear that the "thing" will return. I don't want to remind them of their ordeal; to risk rekindling the psychological conditions that could spark a replay of the whole business. Hence, the long delay--up to a year in some cases-before getting back to them.

Now and then, however, my worries about re-infestation go up in smoke. A post-therapeutic phone call renders me speechless: "What ghost? We never had ghosts!" they assert. The things we study can never fit into the scheme of everyday life. After sufficient time has elapsed, some people begin to doubt that anything out of the ordinary happened to them: they block-out all distressing memories of it. Parapsychology labels such denials, "Retrocognitive Dissonance." It's not amnesia it's rationalization. It is the ability to forget selective unpleasant experiences; a process beneficial to their mental well-being.

So when they say, "I'll believe in ghosts when I see one with my own eyes," I tell them, "Even if you *saw* one you might not believe it. Oh, maybe for a brief moment...but then you'd rationalize it away." Jung called it, "A symptom of the primitive fear of ghosts." Adding, "Even educated people who should know better often advance the most nonsensical arguments, tie themselves in knots and deny the evidence of their own eyes ... because what they have witnessed and corroborated is nevertheless impossible--as though anyone knew exactly what is impossible and what is not! "It is a psychological rule that the brighter the light, the blacker the shadow; in other words, the more rationalistic we are in our conscious minds, the more alive becomes the spectral world of the unconscious." (From the Forward to Fanny Moser's *Ghosts: False Belief or True?* Baden, 1950.)

In the end, many of my clients make the same psychological break from me and from their problems as do patients when psychotherapy has run its course: "How will I know when I'm cured?" they ask their doctor. "When the time comes that you begin to wonder why you're seeing a therapist at all, I'll know you're on the mend."

Chapter 5: The Haunters and the Hunter

Glendower: I can call spirits from the vastly deep.
Hotspur: Why, so can I, or so can any man; But will they come when you do call for them?
(Shakespeare, *Henry IV)*

The Psi Session

During his nearly ninety years on earth Carl Jung, father of Analytical Psychology and protege of Sigmund Freud, experienced many what he called, "Incidents of active imagination..." His writings both before he achieved fame as a theoretician and afterwards seemed to underline the reality of ghosts and hauntings. In theory and practice Jung was about as far from being an Occultist as a person could be; but he did lend a kind of official sanction to its offspring, Parapsychology, long before it was accepted into the family of conventional sciences.
In *Memories, Dreams and Reflections,* a collection of his works edited by Aniela Jaffe (Vintage Books, New York, 1965), Jung said:

> *"Parapsychology holds it to be a scientifically valid proof of an afterlife that the dead manifest themselves--either as ghosts or through a medium--and communicate things which they alone could possibly know. But even though there do exist such well-documented cases, the question remains whether the ghost or the voice is identical with the dead person or is a psychic projection and whether the things said really derive from the deceased or from knowledge which may be present in the unconscious."*

We are no closer to finding, unequivocally, the source of the things said by ghosts and their alleged interpreters, the mediums, today than they were in Jung's time; but one thing is crystal clear: the mind can and often does create projections more astounding than the spoken word. Jung called them "exteriorized unconscious complexes," and the visual and auditory effects they produce, "catalytic exteriorizations." Although in later life he did soften his position somewhat, he rejected the idea that any of it originated in the "spirit world."

Earlier I established that the initiating source of the hubbub is the Agent. He's the "trigger man." At its center stands the Focus, the person around whom psychic disturbances seem to revolve. No phenomena can be created or projected without a living subconscious mind to act as a catalyst. I think the Focus is that catalyst; if not in all, certainly in the majority of cases. Logically, then, no one could know more about the identity and underlying purpose of the Agent than the person whose mind is being used in this manner. That information lies beneath consciousness.

When misidentified natural events, mental aberration, and fraud cannot account for a haunting; when there's no reason to suspect that it's the consequence of Random RSPK (the Poltergeist), or the Psychic Imprint; and where the residents of the troubled house request that I get rid of it (and in my judgment it appears feasible to do so)–my objective is to reach the focal person's subliminal level of consciousness. I'm after clues to the identity of the "ghost," the reason for its presence, and the chance to plant a strong suggestion--a "psychological block" against the mind being used this way. (Experiments during the late 1980s at the Mind Science Foundation in Texas confirm that blocks create an effective and lasting mental shield against psychic phenomena.)

To attempt these lofty goals I use the methods I've developed over the years: a system designed to stimulate subconscious responses. I set up the atmosphere of a "Seance" by borrowing a page from the late nineteenth and early twentieth centuries when seances (French for a sitting) and table tipping were the favorite parlor games in the Occidental world. I light candles, play mood producing music, burn incense, and prefer a night with lightning and a howling wind.

There's no doubt that the right atmosphere helps. Theoretically, I could summon the "dead" in a brightly lit room in the middle of the day. But movies, television, and novels have given a picture of the ceremony that people expect to see. Because it works, it's what I give them!

The Psi Session (my term for a seance) stems mainly from two sources: from the highly specialized entertainment vehicle known in mentalism circles as, "The Ghost Show"; and, from "sitter groups": small groups of sober (and not-so-sober) seance experimenters who sit in dimly lit rooms and try to contact the spirit world.

As a part of their routine, professional mentalists and group members *pretend* to be in control of psychic powers. Curiously, when strong enough emphasis is placed on the supernatural, genuine effects come about. Psychokinesis with ensuing Poltergeist-like displays are often observed. Trance mediumship (where one of the sitters falls into a half-conscious state and utters meaningful dialogue) is another.

Anyone familiar with the "Philip" experiment knows that it's possible to will a "ghost" into existence. In the 1970s, members of the Toronto Society for Psychical Research, under the direction of A.R.G. Owen, created a being identified as Philip. The imaginary ghost was supplied with a whole set of characteristics and made-up family history. He "existed" for over a year by an act of concentrated collective will. Though there was never the slightest doubt that Philip was an imaginary construct, he did mystify the group by rapping out details of his make-believe life. (Source: *Conjuring Up Philip*, by Iris M. Owen and Margaret Sparrow, Harper & Row.)

The Psi Session is a marvelous tool. For reasons explained later, it is applied to suspected Genuine Haunting Entity and Directed RSPK cases only, which account for only a small number of the total I research. Since the impressionable and unsophisticated are strongly affected by it, it must be conducted with great care. To the uninformed the seance is an almost magical performance. As with all "mind miracles," however, the pronouncements of the seance like those of the Ouija board should never be blindly accepted.

The Psi Session has all the trappings of seances performed a hundred years ago, save one: notwithstanding the case I call "The Wailing," I do not normally employ the services of a physical or mental medium. Instead, I make the Focus his own medium. I zero in on the center of the disturbance and, by so doing, eliminate the need for the professionally trained sensitive.

One reason I decided to tap the Focus rather than turn to the medium involves the issue of sensitivity. As a rule, psychics are far too sensitive to impressions from the environment and from the people around them. They're adept at getting information about the dead; but they're also proficient in uncovering facts about nearly everything else in and out of sight. They tend to pick up intuitive material that has little or nothing to do with the investigation. Misidentified perceptions add no end of confusion to a situation already rife with it.

Furthermore, whatever the source of the haunting, entity messages seem to be limited by the extent of the medium's knowledge and vocabulary. Often their personality and presuppositions negatively influence the communication. They are, after all, strangers on the scene. Not so the Focus. Because of his or her direct involvement they're apt to produce information related to it.

More than a few researchers have used trance mediums to help them in their pursuit of spirit mischief-makers. It's interesting that in nearly every case the culprit turns out to be a discarnate entity or evil spirit according to them. One of the criticisms of mediumship is that information comes telepathically. For my purposes I welcome such information because I'm not particularly concerned with its source. I'm happy to get it regardless of where it comes from so long as it leads to a solution of the case.

A legitimate objection concerns fraud. To be fair, most mediums are a mixture of ability and deception--deception to the extent that they probe for telltale clues. Of course, parapsychologists know that fraud occurs. But I believe it's an accusation unfairly leveled at mediums as a group. In any event, in those situations where the Psi Session can be employed I have a high rate of success tapping the more direct and accurate source of information--the Focus.

Breaking the Mind Barrier

To put it simply, my goal is to reach the source of the disturbance. One method allows the mind, or some outside force using the mind, to create movements called "automatisms." Automatisms affect the motion of the pendulum, the hand as it writes, and the rapping sounds made by a rocking table. Without these tools I'd be groping as well as sitting in the dark. Psychologists call the process "Ideomotor Signalling." Hypnotists call it "Calibration" and use it to tune in on involuntarily produced, non-vocal responses. For instance, to judge the depth of their induced sleep, the hypnotist will tell his subject to signal "Yes" to questions by lifting or bending the right forefinger and "No" by crooking the left. They nearly always comply.

Aware of it or not, mentalists and psychic readers employ self-sustaining automatisms in their acts. They (and to some extent, mediums) use it to get "silent" reactions during "fishing expeditions." They watch for subtle visual, or tactile feedback. By asking "Yes" and "No" questions only they're able to observe more than just the spoken word. All they need do is watch their subject closely while they answer. They're looking for the little giveaway actions and mannerisms that accompany and distinguish a "Yes" from a "No."

It's a lot easier than it sounds. For example, some will tighten or extend jaw muscles during "No" responses--relaxing these muscles for "Yes." Others turn blush-red in the face or neck when answering personal questions; then promptly lose the color when the answer is "No." A few stare upward or downward; unconsciously tilt their head one way or the other, or will without thinking give away the answer by nodding or shaking it.

Mentalists and readers become adept at picking up involuntary hints. Skillfully, they move from one supposition to the next, looking for non-verbal clues. They build an impressive body of "pre-knowledge" as they go, leading one to believe that they're truly mystical when, in fact, the bulk of the "right-on" material came from the subject's actions. I don't mean to slight psychics. Many are genuinely gifted. My only purpose is to point out the existence of automatic functioning and the often overlooked skill of interpreting non-verbal, involuntary feedback. Automatisms are activities performed without conscious direction. They occur when a powerful enough need exists to override conscious barriers. If you've ever been hypnotized you know that the body can respond to suggestions automatically: you don't have to think about it, you just do it! Although I am a Clinical Hypnotherapist, the Psi Session is not an exercise in hypnotism. It isn't necessary to induce a trance in a person to get them to operate the devices of the seance. It only requires their willingness to follow directions; automatism takes over from there. Like the crystal ball and magic mirror of the seer, these tools access the recesses of the mind. They amplify tiny motor signals from the subconscious and convert them into meaningful messages.

The pendulum and automatic writing have a long and colorful history in necromancy. The original pendulum may have been used in the days of the Roman Empire; similar devices to the Ouija board, which is itself a form of automatic writing, were known as far back as the sixth-century B.C. Table tipping or tilting is of more recent vintage but is so cumbersome that I only use it as a warmup exercise.

Pendulums are the simplest devices imaginable, consisting of a string or light chain with a weight (a key or ring) attached. With the elbow on a flat surface, forearm upright, and wrist bent, the operator holds the end of the string between thumb and forefinger allowing the weighted end to dangle. Its length should allow the weight to swing freely just above the surface of the table. To communicate, we develop a code from the three natural motions: back and forth along a horizontal line; up and down vertically; or around in a circular motion. For the majority, a vertical path corresponding to nodding the head up and down, will mean "Yes"; horizontal, "No"; and a clockwise, or counterclockwise movement, "Unknown."

A simple test involves the questions, "Am I a human being?" "Do I have two heads?" "What is the meaning of life?" To which each of the three possible replies should follow: Yes, No, and Unknown. In a few minutes the pendulum begins to follow a definite course--moving as though driven by an unseen intelligence. I introduce my clients to the "magic" of the pendulum so they can see for themselves the concept of action without apparent causation. It's always amazing to them that this realm of knowledge can be tapped so easily.

The Ouija is almost as well known as Parker Brothers' other board game, Monopoly. The device permits one or more operators to select letters, numbers, or the words Yes/No/Unknown by pointing to them with a small three-legged tripod... Like the pendulum, the process is ostensibly an unconscious one, unless someone purposely forces replies. In 1892, Ouija's inventor, William Fuld, patented it as a "talking-board" device. Talking-boards are a primitive form of automatic writing, a practice that first caught my attention after I read the late Dr. Anita D. Muhl's book *Automatic Writing: An Approach to the Unconscious* (Garrett Publications, New York, 1930; revised and republished shortly after her death, in 1960).

Dr. Muhl, a psychotherapist, used automatic writing together with other more orthodox approaches in the treatment of emotional disorders. Often she was able to get at unconscious processes more quickly when her patients "wrote" than when they were exposed to standard therapeutic methods. Muhl believed in the use of subliminal writing, drawing and other apparatuses. She was careful to point out the potential hazards: chiefly the increasing tendency for the user to withdraw from reality and neglect routine life in favor of the more fascinating aspects of the experience. The chance to talk to deceased loved ones; to have a "friend and confidant" always there to share fears, hopes, and dreams are attractive prospects for some.

Of course, automatisms are not magical. They're merely the driving force behind the pendulum and planchette. They occur only when the conscious mind is temporarily side-tracked. If they were magical we could expect to get responses without the aid of an in-the flesh operator: the pendulum would swing without being held; the tripod would spell-out messages on its own. Without the human touch, however, no movement, let alone intelligent response is possible.

Anyone who doodles or is a somnambulist, that is, walks or talks in their sleep, can become an automatist--which means that nearly everyone has the ability. Dr. Muhl said that over ninety percent of her patients--people who came from every walk of life--could do it. It's as easy as sitting down to write a letter, except that the automatist allows his or her subconscious mind to do the work. Muhl believed that dissociation (when one or more parts of the mind behave independently of the rest) existed during automatic writing. Her subject was, no doubt, in a state of altered consciousness; and the messages: direct communications from the subconscious mind.

One of the shortcomings of automatic writing is the fact that most have to wait for a time before their hand moves, much less writes anything. A few need more than one session; some never develop the technique. The Ouija is the most popular talking-board ever devised, but slow at bringing in messages. On top of that, occasionally an alleged entity will be unable to read or write, which, of course, renders the thing useless. Somewhere between the Ouija board and automatic handwriting lies the writing planchette. The planchette, French for "little plank," has served as a learning aid for would be automatists since the 1800s. It is to automatic writing what training wheels are to the fledgling bike rider.

Similar to the Ouija the writing planchette is a tripod, too, except at its apex a pencil or pen serves as one of its legs. The device rests on a block of newsprint paper (at least 24 by 36 inches), while the fingertips rest lightly on its flat top. In this manner it writes-out messages in words or ideas expressed in drawings. Pencil drawings are intriguing and, as might be expected, take longer to analyze than do word messages. Whether the "spirit" is illiterate or not, it seems only natural that the subconscious would express itself in drawings, since thoughts like dreams are believed to reside in symbolic form beneath consciousness.

Typically, four to six participants are seated around a table illuminated by candlelight. Even though I use the scatter gun approach (multiple sitters joining-in) my aim is directed at only one: the Focus. During this segment I repeat questions left unanswered earlier. I expand them to include those that need more than a simple "Yes" or "No" answer. My goal remains twofold: to uncover who or what is responsible for the haunting, and to learn what I can do to end it. The tripod "comes alive." It gains speed, then races across the paper. Or it moves at a leisurely pace, writing with deliberation. At first, only straight lines, circles, spirals, figure eights, and scrawls appear. When words come they may be in tiny little letters or take up an entire page. Often they're written backwards, upside down, or in combinations of both. Sentences may flow around the edges of the paper; or words stacked in columnar form, oriental style. In the beginning they're nonsensical, reminiscent of the hodgepodge transcriptions of the EVP buffs, Jurgenson and Raudive.

Spelling suffers. Words are written phonetically, or they're combinations of cursive and printed letters. Numbers are inserted to replace syllables, like 1-derful, 2-night, and 4-tunately. Shortcuts are taken: t's not crossed; i's and j's seldom dotted; vowels--a's, e's, o's, and u's-- difficult to tell apart. It's all terribly clever and reminds one of dreams (the way they're symbolically expressed). Trying to analyze planchette messages is like interpreting dreams, an observation that did not escape Carl Jung. In his paper, "On the Psychology and Pathology of So-called Occult Phenomena" (reproduced in *Psychology and the Occult,* Princeton University Press, 1977), Jung said:

> *"In numerous experiments .....I have noticed, usually at the start of the mental phenomena, a relatively large number of completely meaningless words often only senseless jumbles of letters. Later all sorts of absurdities are produced, words or whole sentences with*

> *the letters transposed all higgledy, piggledy or arranged in reverse order, like mirror-writing. The appearance of a letter or word brings a new suggestion; involuntarily some kind of association tacks on to it.... The recognition of this automatism again forms a fruitful suggestion since at this point a feeling of strangeness invariably arises, if it was not already present in the pure motor automatism. The question 'Who is doing this?' 'Who is speaking?' acts as a suggestion for synthesizing the unconscious personality, which as a rule is not long in coming..."*

Although we understand the mechanics of automatism, there's no actual proof of where the words come from. Until there is it would be a mistake I think to rule out the supernatural. There's always the possibility, if only in theory, that messages originating in the spirit world bypass consciousness on their way to the central nervous system, and finally, to the muscles of the arm and hand.

Nothing points to the existence of an independent subconscious mind so much as the appearance of automatisms during the séance. To the uninformed these self-sustaining actions are easily mistaken for supernatural, or superhuman feats. For that reason it's important to discuss further the theory behind them, lest they acquire an undeserved mystique.

Robot-like functioning is likely to occur when a behavior is so ingrained that we no longer give it any thought. When I make the round-trip from home to work and back again, I often find myself going into and out of such a mode. As long as there are no obstacles in my path my mind is free to wander to things unrelated to driving a car. My body, having navigated the trip to work and back countless times, took over. When some damn fool crosses in front of me, I come out of my semi-trance long enough to hit my brakes, honk my horn, and holler a choice expletive or two.

Back in my youth, before I could afford automatic transmission, I had to shift for myself. Although complicated maneuvers, once you got the hang of them engaging the clutch and moving the gear lever in unison were functions never given a first, let alone a second thought. Some people unconsciously hum. Some folks doodle. Pianists and typists experience it every time they sit at the keyboard. It's perfectly normal behavior; we all do it. Even so, psychologists know that if automatic functioning is carried to an extreme another distinct individuality might surface and take control of our actions.

Dissociation is a state in which a portion of the mind separates from the part not in contact with the outside world and operates on its own. Catholic priest and psychotherapist, Adam Crabtree, *(Multiple Man)* talks about dissociation as it relates to automatic writing and multiple personality. Crabtree tells us that not all of these cases have a psychological solution. Some may be the upshot of diabolic and discarnate influences, while others may be caused by projections from living human minds.

We know from the study of human psychology and physiology that our hands are in a constant state of subtle movement caused, to a large degree, by our thoughts and emotions. These disciplines tell us that automatisms can be induced fairly easily in the average person by verbal and tactile (touch) suggestion. When used in the Psi Session the process is similar to hypnotism, except during my experiments the trance is only partial and the area of influence is designed to extend only to the arm and the hand.

We've all seen movies in which a Svengali-like character dangles a glittering gem, or spins a spiral in front of his subject to induce hypnotic somnolence. (Dracula had only to gaze at his victims to achieve it.) My charges pay extremely close attention to the devices during the séance often becoming so absorbed in them they slip into trance. The principles of verbal and tactile suggestion are applied: I tell them that the planchette *WILL* answer our questions = *verbal suggestion;* then, using the sense of touch to reinforce the spoken word, I take each participant's hand in mine for just a few seconds = *tactile suggestion.* Both are effective induction techniques. The enhanced atmosphere and their eager expectations tend to create the ideal conditions for light trance. Tactile stimulation directed to a specific area, such as the hand, energize the pendulum and planchette into action.

The Discernment of Spirits

Clerics say a prayer for the discernment of spirits so that they may know what they're dealing with. If demonic, back to hell they go! If human, the poor soul is released to its eternal rest. Before going very far with the seance I demand that the communicator identify itself (or at least tell me who it thinks it is). My concern is not so much that I've conjured-up something malevolent as it is that I've plumbed an area of the mind better left alone. It's not only pointless to continue the contact if I've reached some uninvolved earth-bound spirit or inhuman creature (if such things exist), it could prove to be dangerous as well. Mind dredging is a tricky operation for the most skilled therapist. When threatening words are recorded or at the first sign of acute stress I stop the ceremony. Unless the source can be identified, and in my opinion poses no threat, we won't go on--no matter the temptation to continue.

Between the 1850s and the turn of the century table-tilting was the rage. When used in the Psi Session it's an exciting if not rewarding test of PK. With a lightweight table resting on a smooth surface, I ask the sitters to place their hands lightly upon the surface; palms down, fingers spread apart in such a way that their pinkies are touching. As long as they continue to touch, an uninterrupted contact, or flow is maintained.

I ask if there's an unseen presence that wishes to speak. Sometimes the table will respond immediately; normally, ten or more minutes pass before it does. During the performance, short, irregular undulations are followed by a strong lurching or rocking action. Two legs rise off the floor; the opposite pair remain firmly planted. When it rocks back, down they come with a force that produces a resounding thump. I apply a code to the thumps, or raps, which serve as the vehicle of communication.

It's a long proposition and my least favored way of getting messages.

Sometimes it's uncommunicative yet far from unresponsive. Instead of rapping out messages the table levitates, three legs off the floor, while doing a pirouette on the fourth. I'd have to see it myself, but I've been told that on rare occasions it lifts off the floor completely. No less impressive is the galloping table. Take it from me, when otherwise respectable pieces of furniture try to run away on all fours, sitters' chairs and everything else in their path go flying. And what are my clients doing during the craziness? Resembling reluctant riders on a runaway horse they're holding on by their fingertips. In the middle of a tough case, the sight of a galloping table and its frantic operatives running to keep up with it helps break the tension-filled spell.

It's an impressive demonstration. Nevertheless, I don't get hung-up on the theatrics of acrobatic tables. Its value in solving the mystery, which is after all the reason I'm there, is minimal. Tables move for the same reason pendulums and planchettes do: the mechanics of motor automatism, with a jolt of RSPK thrown in for good measure.

Case File: The Haunted Restaurant

The apparition of a young woman had haunted a Florida restaurant for over six decades. The first instance on record was in 1932, when a group of construction workers participating in the buildings' renovation saw her glide through their midst, then pass silently through the wall of the kitchen. From then on patrons and employees said they saw the ghost on many occasions.

In 1992, the building housed a restaurant and a pub occupied the top floor. A bartender and two waitresses told me that equipment would start up on its own: switches engage with no one near the controls; and cash registers and kitchen appliances cut-out mysteriously. Now and again patrons and employees would catch a glimpse of a mysterious form: a young woman reflected in the glass windows of the dining room.

During a seance we had a rather peculiar chat with a young lady: peculiar when you consider she'd been dead for sixty years. Because the ghost could neither read nor write well it took a good deal of time to establish contact. When we did, she "spoke" to us through the writing planchette in a kind of hieroglyphics: words in line drawings. An unknown assailant had brutally raped and strangled the woman around the time of the depression (1928-1930). She'd been buried by her attacker in a crude grave somewhere under or near where the building now stood. Bound to the site all these years her spirit was trying desperately to attract attention to its plight. Although I searched the local library archives and the newspaper morgue, I found nothing that could confirm the story.

We instructed her to move on; but sometimes they refuse to go! Still, I'm happy to report that our visit wasn't wasted. It comforted those who spent their nights in the building to know that she meant them no harm. In 1995, the old building was razed. The site was paved and turned into a parking lot; and that was the end of the haunting. According to a rumor that circulated the area, human remains were removed during excavation, then interred in hallowed ground. I'm afraid that part of the story had all the makings of a "friend of a friend tale," since it could never be verified.

Reframing the Mind

I don't always succeed, but the main goal of the séance is to reach the subliminal level of consciousness and provide a block, or "reframe" the mind against being abused by the offending "ghost." Reframing is the term hypnotists use to describe a change in a particular pattern of behavior. Hypnotherapists use it to help their patients kick bad habits: smoking, drinking, etc., by working directly with the subconscious without using the conscious mind as an intermediary.

Using suggestion, I attempt to change the patterns that cause phenomena to be exteriorized. Where the victims believe in the paranormal nature of their haunting I use a simple demand such as: "The disturbance that is disrupting this household is no longer permitted to manifest!" Where they're downright adamant about it being a spirit invader I'll go a step further and say something like: "We ask that you help us help you so that you will not remain earth-bound, and so that you will no longer disturb those in this home. You must seek your proper place on the other side of life. You must not remain where you are unwelcome. If it is your genuine desire to move on to your reward, all you need do is sincerely ask for help to guide you there and you will receive it."

There are times when, for whatever reason, the source of the haunting will be insufficiently moved to cease and desist without the use of stronger language. When that happens I fall back on words borrowed from the formal exorcism, or from the house blessing ritual to get the job done:
YOU ARE FORBIDDEN TO REMAIN IN THIS HOUSE! I COMMAND YOU TO CEASE YOUR PERSECUTION OF THIS FAMILY! DEPART AND, HARMING NO ONE, RETURN AT ONCE TO THE PLACE APPOINTED YOU, AND REMAIN THERE.

The reader should keep in mind that most of my work is carried out without the need for admonitions, biblical exhortations, or séances... On the other hand, when they are used the reason they work has a lot more to do with the impression I make on my clients than the one I make on their "ghosts." The belief that I'm "the expert" instills in them an anticipation of success. It's an emotional placebo, a self-fulfilling prophecy of my victory over the nether world. Combine that with the highly suggestive nature of the Psi Session and you can understand how Reframing might be an effective method of "cleaning" haunted houses.

It's nothing less than post-hypnotic suggestion. The equation representing the formal induction of hypnosis is: misdirected attention, plus belief, plus expectation and imagination, equal hypnosis. Anticipation and the right atmosphere tend to produce not only useful information through automatisms, but a climate ripe for the command that the haunting will end. Whether acted out by medium, stage performer, or Ghost Detective, there's little doubt that when positive results are gotten they're due to the effects of a suggestion to the subconscious mind.

Hypnosis (heightened susceptibility to suggestion) is a short-cut to the unconscious. It is an altered mental condition, and, for purposes of the Psi Session, the most useful thing about it is the elimination of the individual's active belief system. The conscious mind is ever aware of its own self-imposed limitations. The subconscious, a transition zone between the unconscious and consciousness, has no such boundaries; is open to suggestion and will try just about anything if instructed properly.

Yet my clients are never in any danger during the ceremony. Sir Philip Sidney said, "The truly valiant dare everything but doing anybody an injury." The mind can protect even deeply entranced subjects by rejecting harmful or immoral instructions. Throughout the duration of the Psi Session the individual is in an awake state, fully aware of what is taking place and active to the extent that he or she is able to manipulate the psychic devices. It is a condition in which they show an amazing ability to focus on the given task.

I'm not the first to work with altered consciousness with the aim of expelling ghosts. Hypnotism, or more accurately its forerunner, Mesmerism, was used in the early nineteenth-century--years before the founding of Spiritualism--to establish contact with "the dead" before removing their negative influence. Supposedly, low spirits lacked intelligence and were suggestive to the will of the mesmerizer. Researcher and author, Frank Podmore, briefly discussed its uses in *Mesmerism and Christian Science: A Short History of Mental Healing* (George W. Jacobs & Co., Philadelphia, 1909). Perhaps because such practices ran afoul of the Church and popular sentiment, they were discouraged.

Over 150 years ago German psychologist Justinus Kerner used hypnosis (then called "magnetic treatments") to expel ghosts. Many thought his methods were psychological tricks since they relied heavily upon suggestion; yet they were nearly always beneficial to his patients. Even Freud used hypnosis early in his career. According to A.A. Brill *(Freud's Contribution To Psychiatry,* W...W. Norton & Co., New York, 1944) ".....hypnotism was absolutely indispensable for the cathartic treatment" [treatment that brings repressed material through the subconscious on its way to consciousness]. In time he became dissatisfied with the therapeutic results of catharsis based on hypnotism. Though the results were sometimes striking, they were short-lived. Those who were hypnotizable invariably suffered setbacks after only a brief period of relief. Hypnosis, in Freud's case, turned out to be only a stepping stone on his way to psychoanalysis.

My "cures" using the power of suggestion are usually long-lasting. It has nothing to do with me but rather to the fact that early life trauma (the suspected cause of a number of neuroses) is not usually connected with hauntings. I'm convinced that anxieties present in the Focus create the energy that propels phenomena. When I break the ghostly cycle, he or she unconsciously re-directs these emotions towards more constructive, or at least less destructive, ends; and the haunting comes to a quick and permanent end. It's as simple as that!

Though the results are similar, what I do during the Psi Session is the reverse of what the faith healer does for sufferers of mind-created ailments. I promote a psychological process to *block* the spectacle of the ghost. I command the invading entity to leave forthwith. Invoking the name of God, or other deity, the faith healer removes a block. He commands that they stand and walk, or see and hear again. If the ailment is not rooted in disease, nerve damage, broken bone, etc., and the victim's belief system is strong enough there's a good chance the psychological block will be removed. (I'm not overlooking Lourdes and La Salette, but genuine miracles are so rare and psychosomatic illnesses so common.)

In other words, short of miraculous intervention, the truly sick and injured probably can't be cured that way. In the past, after a particularly tough case, the thought has occurred to me that the truly haunted may be just as resistant or perhaps immune to my brand of hyper-suggestion: that an obstinate energy force was holding on to them. Whether this force was devilish, discarnate, or divine I do not know.

The Framework of Their Belief System

Why do I direct questions to a "spirit" when I'm more than a little skeptical of its existence? I do it because from a therapeutic standpoint it hardly matters if it's Aunt Mamie returned from the dead or a thought-form created by her niece or nephew. The course of treatment I call the Psi Session works in either case. Dr. Walter Franklin Prince, for many years the research director of the American Society for Psychical Research, was one of the first to "professionally humor" his patients. Prince believed he was practicing a form of beneficial suggestion by combining elements of exorcism and psychotherapy in his work. In "Two Cures of Paranoia by Experimental Appeals to Purported Obsessing Spirits," he wrote:

> *"My acknowledgment that the imagined obsessions might be actual ones, my agreement that the patient might be right in his opinion [that he was being plagued by spirits] put him into a favorable state of mind, a state of satisfaction with the admission, in which my exhortation to the imagined spirits to cease their mischief could act as suggestions operating in the patient."*

As far as I had the ability to determine, none of my clients were suffering from paranoia--or any other mental aberration--when they entered the séance room. Nevertheless, the principle of using suggestion to stop the "mischief" seems to have worked for them, too. Adam Crabtree, the Canadian priest-turned-psychotherapists, is another proponent of treating disorders by "going along" with the patient's delusions of obsessing and possessing entities. In his book *Multiple Man* (Praeger, New York, 1985) he tells us that even if these forces do not exist, sharing the idea of their reality can be of enormous therapeutic benefit.

To be of any help to my clients I must work on the level of their understanding. Those who have strong religious ideas and those who have a mystical side to their nature are prone to turn to the supernatural for answers. I don't argue with them. I work within the framework of their belief system.

Of course, it's best not to play the "ghost game" when there's a natural explanation. Otherwise I may, by my silence, give tacit acceptance to the notion that they're haunted. And as far as Agents and Focuses are concerned, it's difficult enough to dissuade families from the supernatural, much less convince them that one of their own is producing the phenomena. I'd run the risk of having the poor soul declared "possessed," if I did. Perhaps the best feature of what I do is that it brings the family together as a group to "heal" their problems. Whether you believe the invading entity is the essence of someone who once lived, or mind-vented frustration, hauntings are the expression of a need. The investigation provides the means to learn what that need is and, when possible, bring it to realization.

Case File: A Poltergeist From Neptune

"The test of a religion or philosophy is the number of things it can explain."

(Emerson)

Maybe this guy and I were kindred souls, but my first thought was to lose his phone number. Even I draw the line somewhere, and when you start hearing from "little green men" it's time to get out the chalk. Except this case had a peculiar hook. The whole family, it turned out, was either a study in the psychology of the deluded, or a lesson in the banality of evil.

I quickly discovered that Aubry Wilton (a fictitious name for our forty-six-year-old caller) was not your everyday contactee. Aubry was employed as an aerospace engineer at one of the nation's largest companies. His words were chosen carefully and spoken in a manner that suggested a scientific background. He struck me as a calm, levelheaded, altogether credible witness. His demeanor changed slightly when he boasted that earlier that year he'd come back to Christianity for, as he put it, "The perfect protection of our Lord." Unlike those who in middle age seek the subtle comforts of religion, Aubry Wilton was the epitome of the zealot. No Sunday go-to-church, yawning postulant was he. Faith without great passion was no faith at all. Aubry yearned for the fervor and traditionalism of the old-time religion. Living on the fringe of the bible belt he had no difficulty finding them.

"Three years ago," he told me, "at a moment of great despair, I became a born-again Christian." Now, in the summer of 1988, he had more reason to rejoice. He was convinced that the renewal of his faith had saved his family from evil abominations. His less sophisticated wife, whom I'll call Maybelle, had had strong church ties in her youth. At thirty-five, however, she'd come to believe that the outwardly pious were hypocrites. To Aubry's regret, dabbling in the Black Arts had robbed Maybelle of her religious convictions. "She's not against God. Far from it," he explained. "She has no use for organized religion anymore ... that's all."

The Wilton's have a daughter, Julie, thirteen at the time, and a son, Toby, who was ten. In July 1988, Aubry told me that he and the kids were "worried about Maybelle's visions": "May's a clairvoyant. She dreams about things and they come to pass. She's been 'seeing' some poor soul about to be buried alive; dreams about it nearly every night. It's our number one concern." His outer space connection came to him in a voice channeled through Maybelle's sleeping body, he said. Unfortunately, what Aubry didn't mention was the fact that *he* was the only one who heard it. What was worse, he hadn't bothered to tell Maybelle or the children about this visitor from our neighbor planet. Neptunians weren't the Wilton's only problem. Aubry glossed over the others with the comment, "There were difficulties in the past. I took care of them, so we needn't concern ourselves with them now." Nevertheless, I persisted:

Andrew: "What kind of difficulties?"

Aubry: "The voice and the dreams didn't start until after I exorcized the house. I'm a believing Christian, okay? So I know that an abomination is not involved here."

Andrew: "Wait a minute. Go back to the beginning-to before the voices and dreams. What was it you exorcized?"

Aubry: "It has *nothing* to do with what's going on now!"

Andrew: "I'd like to hear about it anyway ... if you don't mind."

Aubry: "Well, it started six or eight months ago. We heard thumping noises in the walls and scratching sounds and the sound of pennies crashing against the mirrors. There was a black image that smelled of sulfur. It pulled the covers off our beds while we were in them. "I was just beginning to get involved with the (born-again) movement. If you think I'm a nut now you should have seen me then. Well, I asked for help from the minister, and he gave me a prayer to say to expel evil. He told me that Christ would not permit abominations to exist in the same place with a believing Christian. He was right! Everything stopped after I said the prayers ... except that night we heard loud banging sounds on the walls. But everything else was gone. Okay?"

Andrew: "When did the new things start?"

Aubry: "The next night. We went to sleep at midnight with the pounding going on. Then sometime around two-thirty something woke me up. I don't know what it was, but I rolled over toward May and she started speaking to me in a deep voice. Her eyes and her mouth were closed but the voice came out of her. That's when I met Morla."

Andrew: (Ignoring Morla for the moment) "What makes you so certain the abominations, as you call them, are gone now?"

Aubry: "I'm absolutely certain they're gone. They can't abide in the home of a true Christian."

When Aubry Wilton first mentioned his extraterrestrial contact and his wife's nightmares, I mentally wrote the case off. His Neptunian and Maybelle's dream were not grounds for an investigation. After hearing more, I was drawn to the fact that the disturbance had been transformed overnight, apparently in response to his new-found belief system. In the beginning, typical Infestation phenomena (recurrent spontaneous psychokinesis) followed by the sight of an ethereal form was endured by all four family members. After his attempt at religious expulsion a type of channeling seemed to be taking place, in addition to a strange dream-one reminiscent of Edgar Allan Poe's tale of premature burial.

And there was something else. Aubry told me that on three occasions Maybelle awoke from her nightmare to find blood on her hands. Each time her body was searched for wounds, scratches or abrasions, but there were none. The blood was neither hers nor his, but it did appear to be blood. The circumstances were totally unique. Two days later I was interviewing the Wiltons in their home.

(From the tape) "It's 7:00 p.m., August 1, 1988. We're in the Wilton's living room":

Aubry: "May's clairvoyance doesn't have any relationship to Christianity, you know. This is some source of power that I don't know how to deal with."

Maybelle: "My biggest problem is the nightmares keep me from getting my sleep. I see this poor woman inside a coffin. There's a window in it so you can see her face. She's knocking on the lid; scratching at it, trying to make noise ... to get out of there before it's too late."

Andrew: "Too late for what?"

Maybelle: "To be buried alive! To suffocate to death. And I hear ..... and (turning toward her husband) he's heard it, too ... like if you get something with wheels on it going down a track and it's real squeaky. I've tried to lift the lid from the inside. But I don't see me as being inside the coffin. I don't see my face when I look through the window. I see eyes. I hear screams. I see a woman's face but I can't make out who it is. I don't see myself in there. I push up on the lid and it opens a little, then falls down on my thumbs. I know I gotta find a way to save her. I told Aubry that he better call that guy back (meaning me) because I need a shrink. I'm going crazy. I've thought that maybe she's on her way to be cremated? I don't know. Maybe that would explain the wheels on a track. It's not the first time I've seen the future." Maybelle related a similar dream from several years before: "I knew I was dreaming, but I knew that the boy was really there in the mine shaft. Only I didn't know who to tell-who to warn because I didn't know his name, or the names of his people. And I didn't know exactly *where* he was."Her eyes filled with tears.

"It's terrible, you know," she went on. "To know that something like that is going to happen but not be able to help the poor person. The dream didn't stop until I read they'd found a young boy in a mine shaft in Germany. He was dead. And now this..."

Aubry: "Those kind of things happen to May all the time."

Andrew: "How long have these dreams been going on?"

Aubry: "This one, off and on since I chased the spirits out of here."

Maybelle: "The pain in my thumbs wakes me up. You know how pain will wake you up? If it's just a dream, why do I have it all the time?"

Aubry: "And why does she have smashed fingertips and thumbs on both hands, like the lid fell on them? And all the blood?"

Maybelle: "I've got blood all over my hands when I wake up."

Andrew: "After you wash the blood off, what do you see on your hands?"

Maybelle: "Nothing. They're not even pink. They're just natural."

Aubry: "There's no blood in the cracks. It washes off completely. It's just like it's light surface area blood."

Aubry: "There's something in here that looks for her when she's not at home. Whatever this thing is, it searches for her. I can sense it."

Andrew: "I'm surprised you said that 'it' searches for her since you seem to be convinced that it can't be a ghost or a spirit."

Aubry: "No, it can't. I call it “it” because I don't know what else to call it. This thing seems to be getting stronger and stronger as it goes along-not only stronger in duration but more variable things are happening. Her fingers and thumbs are getting smashed on the ends. Okay, it's like somebody took a hammer and just went *bunk* and mashed her finger tops."

Maybelle: "It hurts!"

Aubry: "But there's no blood, no bruises. And there's a mark. .. an imprint of a straight line along the five fingers ... a crease that goes away the next day, as though she were lifting up like this (palms upward). And when the coffin lid falls she's got her thumbs in there, too."

Andrew: "Aubry, she said you hear squeaky sounds at night?"

Aubry: "Let me start way back in the beginning. She started hearing this squeaking at intervals, like a rusty caster being pushed: squeak. .. squeak, squeak...squeak, it would go. And I didn't hear it because it wasn't on my side of the bed. But when I went over to her side, I heard it ... but only in the dark. When I turned the light on, no matter where I was or where she was, it would stop! When I turned the light out, immediately it would start. Then it got louder and louder regardless of where I was at the time. "Now a lot of nights I'll sleep in my daughter's room, because I snore so loud I wake up the whole house ... and she sleeps with my wife. And I was in there three nights ago, and I thought I was sleeping good and felt something really shake my foot ... followed by this knock, knock, knock on the wooden book shelves behind me."

Aubry: "I want to ask you a question now. In the article I read about you it said you sometimes use hypnosis in your seances?"

Andrew: "I make use of a light trance state, sometimes."

Aubry: "The trouble with a seance is you open yourself up to the spirit side of things ... which this is not! Okay? This is different than evil spirits."

Andrew: "I'm really interested in why you're so certain of that."

Aubry: "Why am I so sure of it? Because this is a call for *help.* Okay? Evil spirits will try to make you believe they're *helping* you but will misguide you ... even the small spirits. And there are different grades of spirits."

Andrew: "You put them all in the evil category?"

Aubry: "Yeah. There's no good they do, and you know them by their fruits ... and they will deceive you. Okay? They will lead you down the golden path and make you believe they're doing something good for you; but they're paying their price for where they're at. And this (what Maybelle was going through) is not that."

Andrew: "Everything that's happened to your family since you invoked the prayers to remove evil influences is in no way connected with spirits ... evil or otherwise. Is that what you're saying?"

Aubry: "That's right."

Andrew: "Is this Church dogma, or is it your own opinion?"

Aubry: "No, this is a combined Christian belief from what's in the bible, and it's verifiable experiences with demons. And it's research from Shirley MacLaine's book where she had a couple of experiences. And it is just reading and being knowledgeable about seances ... what they open you up to. In seances, the medium has to use her own force of will or psyche to contact these spirits. Okay? And there's only two ways to go in the spirit medium: there's only good and there's evil in the particular spirit world I'm talking about. "Now there must be another plane of astral projection or something that May is onto that doesn't deal with this, 'cause this is on the positive side. When you go into the seance type area, which is Christianity and science, there's only evil and good in the spirit world. But there's *light* evils and *heavy* evils. Okay? And they all lead to the same thing: they do no good at all!"

Maybelle: "I have no idea what you're talking about!" (She wasn't alone.)

Aubry: "Okay. Okay. Even though this dream is scary, it appears to be ... and I'm not saying it's for sure ... but it appears to be on the positive side someone asking for help in some shape, manner, or form. And May's clairvoyance is reaching out to someone in need of help. This person doesn't realize it's causing us all these problems. Okay? It has to learn *not* to communicate with us and leave us alone. But this person is not a spirit. It's someone alive. It cannot be cast out as an evil spirit. It's not an evil spirit because I am a Christian; I've cast out the demons in this house. Any evil spirit can be cast out in the name of Christ, and it has to go!"

Andrew: "In other words, you believe your wife is having a psychic experience in which she 'sees' a woman ... a woman who is now living and who is about to be buried, or burned alive. Is that right?"

Maybelle: "Yes! Yes! Only I don't know who she is, so I can't warn her!"

Aubry: "That's the only thing it can be is clairvoyance."

Andrew: "When did you cast out spirits in this house?"

Aubry: "Oh ... six, eight months ago."

Andrew: "Tell me, again, what did you do to cast them out."

Aubry: "You have to have two things. You have to be a Christian and a believer, and you have to cast them out in the name of Jesus Christ. There is another method you can use by anointing and walking around ... casting them out if they're in the house, or living in the walls. (Nodding his head, in a half-whisper he added) They do live in the walls of houses."

Andrew: "You told me over the phone that someone in the house saw a demon."

Aubry: "I've seen traces of them. My daughter, Julie (not present during the interview), saw a demon. I've never seen one because, being a Christian, they won't come around me."

Andrew: "What kind of traces?"

Aubry: "It's like a proton trace. You never see a proton or an electron under a microscope, but you see its trail. These things left trails. They left blood all the way from here out to the back door. And the door was open. And these demons don't have any use for doors. They transfer through walls and stuff like that; I don't know how. So the only reason the door was open was for that demon to let us know he'd been there. I have no idea what manner, shape, or form these things take, but when I cast them out, they go ... all the time!"

Andrew: "How long does it last?"

Aubry: "My casting out is only good for two days. But I went to a church service invoking the blood of Christ, and my name was mentioned in the church and that lasted for three weeks."

Let me interject here that the purpose of disclosing this material is not to cast aspersions at anyone's religious beliefs. Having come this far with me the reader must know that I favor getting rid of these forces, whether material or illusory, by whatever means necessary. I have successfully employed religious provocation on a number of occasions. I do not make light of its use as an effective psychological tool. I turned to the "little green men":

Andrew: "Okay. You said you had an extraterrestrial contact. What was that all about?"

Aubry: "Uh, oh! May don't know about this one. I'll be in trouble for this."

Maybelle: "What?"

Aubry: (Visibly shaken) "Yes. Yes. I hope you can understand this because it's crazier than anything before."

Andrew: "I'm interested in a physical description of it and what it said to you."

Aubry: "I don't have a physical description of her."

Maybelle: (Shaking her head) "Why didn't you..."

Aubry: "She's here because May is psychic. This has to do with her psychic ability. Okay? Psychic people have protectors ... guardians, and they're not guardian angels. They're real physical presences. The one she has is a female. Her name is Morla. She comes from the planet Neptune. She can make it here in three minutes and twelve seconds, at the speed of light. She can vary her speed ..... and she's invisible."

Maybelle: "What are you talking about?"

Aubry: "Except when she starts from zero point. When she starts to leave you, you can see a flash of light, which I did out the back window. She says she is not known here, but the reason she's here is for protection. She wouldn't tell me anymore than that, plus some other things. The reason she's here is that her people on the planet are all good. They're a psychic race, but there's an evil race-just as evil as *they* are good. Okay?

Andrew: "Everything you learned about this race came through your wife's body? Is that what you told me on the telephone?"

Aubry: "Yeah. Morla...she said to 'call me Morla.' She said because of my love for Maybelle I am psychically connected to her. That's how she is able to talk to me through her body. When I talk to her, what I'm saying to her is immediately transferred to Neptune through psychic power. Now, I sat and I timed it. I stopped and looked at the clock, and it was exactly three minutes and twelve seconds that that being was there. And I was talking to her through Maybelle's subconscious."

Maybelle: "You were talking to *me?"*

Aubry: "Yeah. Remember when you told me you always felt that there was a guardian looking over you? Everybody thinks that guardian angels are a fairy tale. Right? They may not *be* guardian angels. Okay?"

Andrew: (Trying to find my way out of the "wonderful world of bewilderment") "This is all information that you received from Morla through your wife's body. Right?"

Aubry: "Right. When I was talking to her I was trying to pump her for scientific information, but they wouldn't let her tell me. She'd start to tell me this stuff, then say, No, I'm not allowed to tell you that ...I gotta go now, and she'd leave."

Maybelle: (Growing more confused by the minute) "When was this? When did all this happen?"

Aubry: "Now! It's still happening. And when Morla talks through you it's in monotone; she uses words that you don't even know. Like the word for Vitamin E on Neptune is 'electropod.' And the term, 'turn off your sunlight' means turn off your lamp, here. "Now the first dream she had about the coffin, I was talking to Morla when she had it. I could tell there was something different about the way she spoke and the way Maybelle was sighing in her sleep. I said, “May is having a happening, or something.” All of a sudden I said, “We need help here. Something is happening that I don't know anything about.” Morla said, “It's a happening. It's in the future. It's not happening now. There's nothing I can do for it.” “And boy! Am I in trouble now if Morla is around... I'm not supposed to tell this to anyone. Now, if that's not wilder than UFOs I don't know what is.”

It was wild all right. It was so outlandish that I stared at him in stunned silence. And his poor wife. No doubt this was the first she'd heard of Morla. She sat there bewildered; hands clasped, shaking her head. And the man wasn't finished:

Aubry: “They know all about Christ on Neptune. But the other being that was here, the dark evil one, it was as though Morla was after it ... to get rid of it. You know, to protect us from it.”

Andrew: “What did you think when this, this alien told you they know all about Christ? Did that disturb you?”

Aubry: “Yeah. 'Cause I was checking them out because I didn't believe it. We know all about Christ, but I'm not sure other worlds do. "She said, 'We're on your side, Aubry.' So I put a bible on Maybelle's butt, and I tried to cast Morla out in the name of Jesus Christ. But *she* didn't go anywhere. Okay'? *So she's* not an abomination.”

Andrew: “You knew she wasn't a demon because she didn't leave when you cast her out? Did she stay around for long after that?”

Aubry: “Yeah. She just said, ‘I'm on your side, Aubry.’ When the demons were in the house, I put a bible on May's chest one night and they couldn't get through to her.”

Maybelle: “Was this before my dreams started?”

Aubry: “This was six months ago! So after the demons split we got this evil one who was talking through May's subconscious: this evil Venusian.”

Andrew: “When was this?”

Aubry: “This was after I cast the demons out. I heard him talk through her, so I said

'The heck with you,' and I ran and got my bible and put it on my chest, and I was up on the headboard."

Andrew: "Why did you put the bible on your chest?"

Aubry: "Because that's the word of God. We'd just had the demons and somehow the word, which is absolute truth, will protect you if you have any qualms about not being a full Christian."

Actually, since he had the "perfect protection," I was wondering why he hadn't put the bible on his wife's chest; but I didn't pursue it.

Andrew: "Why do you suppose you and your wife have been targeted by demons?"

Aubry: "*There's no demons!*"

Andrew: "No, I mean why were you bothered by them *before?*"

Aubry: "When you see these things or have these happenings is when you no longer associate with an evil group, turn to Christianity, or come near Christians. Because the evil side doesn't want you on the positive side. They try to scare you back into the fold."

Maybelle: "You mean when I was in that Black Arts group for a short time?"

Aubry: "Now she, through her psychic mind, could see these demons when they were peeking around the corner in the hallway at us."

Maybelle: (Softly) "I didn't know if they were demons, honey."

Aubry: "But you knew *something* was there. Okay?"

Maybelle: "I was scared. Something was there. I don't know what."

Aubry: "One verifiable experience I had was with a knocking demon: a guy that just comes up and knocks around the house. I cast him out and told him to go back to hell where he came from ... and the knocking stopped. Then one night May got up with the same knocking. I didn't cast it out because I wanted to see what happened. It followed her around to the window and into the kitchen and just kept following her and knocking around her."

Andrew: "Do you think there are demons in everyone's house."

Aubry: "No. But she can sense the evil spirits when they're around."

Andrew: "Then why would they be in yours?"

Aubry: "They came here because *she* was getting next to me. I'd had a physical experience with the Holy Spirit."

Maybelle: "She? You mean me?"

Aubry: "Yeah, you!"

Maybelle: "Well, I am close to you. What do you mean?"

Aubry: "No. No. Okay, because you weren't a total born-again Christian. A total turn over into Christianity."

Maybelle: "I'm still not."

Aubry: "I know. And I was. I had a physical experience. At a time when I was destitute, I prayed and a light came out of the sky and it lit me up like a light bulb. And I cried with happiness for a whole year. Because I saw what heaven was like."

Andrew: "How long ago did that take place?"

Aubry: "Three years...May the twenty-sixth. And that was my Christian birthday. I was born again that day."

Andrew: "I'm interested in your thoughts on demons, Mrs. Wilton. You're psychic. Did you ever think there were evil spirits around this house?"

Maybelle: "In this house? No!"

Andrew: (Turning to Aubry) Well, if you think she can sense the spirits and she didn't sense any, and you did..."

Aubry: (Obviously irritated) "Okay. What did you sense the times the knocking was happening and stuff was being thrown around ... during that time?"

Maybelle: "Oh. Oh, I know what you're talking about. Okay. He had brought up the fact about the evil and all this because, you know, I told him about me going to that place where they studied demonology and all that stuff. They were weird people and I didn't want to have anything to do with them. I went to one or two of their meetings, then never went back. But as far as the evil goes, I think ... just listening to him ... he thinks they're reaching out for me. And maybe they are. I don't know."

Andrew: “All right, anything else? Have you told me everything?”

Aubry: “Yeah. I think so.”

Andrew: "Well let me tell you, you have some pretty far out theories here. But what if they're only partially right?”

Aubry: “That’s fine. I'm willing to learn ... praise the Lord.”

Andrew: “Almost everything you've told me tonight fits in nicely with an ordinary haunting. The covers being pulled off, the knocking sounds and all the rest. I'm not convinced it's caused by an evil force or that it's extraterrestrial, though. What if there's just an everyday energy source here that's dying--pardon the pun--to communicate, but you, Aubry, won't give it a chance. The first thing you do is cast it out ... you're so certain that it's evil. Now I'm not saying for sure that it isn't; but I don't think so. If you had called me and told me about everything that was going on except for Morla, I would have told you there wasn't much going on in your house.”

Aubry: “Okay. But what about Morla?”

Andrew: “It's my guess that Morla stems from the same source as the rest of these demonstrations. This energy source--whatever it represents--was forced to change its tactics when you forbade it to manifest in, what you considered to be, an evil form. Let me ask you this, Aubry. What does your religion tell you about space creatures?”

Aubry: “My minister says they don't exist.”

Andrew: "Right! And because there's no religious dogma covering them they were a perfect guise for these things to take. They could go that route because your mind would accept them in that form...that is, once you thought you'd expelled them in their demonic guise through prayer. "You told me yourself that Morla is a material creature, and yet you must know that according to current scientific data there's massive evidence against any kind of life existing on that planet."

Aubry: “It's a paradisal world. According to her it is.”

Andrew: “It's the worst of worlds! A hellish, lifeless place. This thing could no longer show itself to you as a demon, so it had to disguise itself in a way that would be acceptable to your mind. Except for the substitute words for Vitamin E and for lamp, did Morla ever say anything to you that you, yourself, didn't already know?”

He shook his head. But I could see by the look on his face that I was getting nowhere using this line of reasoning. Aubry was positive that the evil presence and Morla were separate manifestations--opposites in fact. Nevertheless, I offered the following:

Andrew: “Now here's what I want you to do. The next time that *evil* Neptunian shows up let him see your bible. Put it on Maybelle. Better yet, show him a crucifix. Maybe the bible you have is not a strong enough icon. If you don't have a cross, make one with two sticks. That'll work, too. I think you're going to have a communications breakdown. “And just to be certain that Morla is who *she* claims to be, give her the same test the next time you hear from her. You see, in the information on demons, these evil messengers are usually pretty stupid beings. That may be why she couldn't answer your test questions. She may not have been bright enough to make something up for you: you being a scientist yourself.”

Aubry: (Bristling, now) "I'm just not afraid! I have no fear of these evils. This evil could be standing there and I'd just laugh at it. Okay?"

Maybelle: “Then why did you put your bible on me?” *(Why had he put it in on himself?* I wondered.)

Aubry: "That was something entirely different. It was happening to you, and you couldn't cast it out on your own. (Turning toward me) I love her and I want to help her. I'd love to help this other person too ..... whoever *she* is."

Andrew: "You mean the woman in the coffin?"

Aubry: "Yeah."

Maybelle: "What about her?"

Andrew: “I'll tell you my opinion; but you won't like it. I think the person in your dream is you. In my experience, when you dream of someone in a tight spot who has no identifiable face, the person usually represents the dreamer.”

Maybelle: “No.”

Andrew: “Don't you think that maybe that person is you? Subconsciously, perhaps, she's trapped in some situation in her mind. You said yourself that you tried to lift the lid from the inside.”

Maybelle: “What about the boy in the mine in Germany? That wasn't me. I couldn't see his face either.”

Aubry: “What about the blood and the crease marks on her hands?”

Andrew: “You're right. I can't explain them.”

Maybelle: "What would a psychiatrist say if he saw the blood and the marks?"

Andrew: "If they were genuine, he'd have to believe. And he'd believe it all the way into the next day. Then rationalization would take over and he'd doubt, and probably even deny, his own senses."

Maybelle: "But why?"

Andrew: "For the reason that what he saw wouldn't fit in with what he knew about the physical world. He'd have to discard it or give up science. He'd have no other choice."

Aubry: "It wouldn't fit his 'fishbowl.'"

Andrew: "I do think that you, Maybelle, may have an extra pinch of psychic ability. You must learn to protect yourself from your visions. You could be trained to do that ... to learn how to block them out so they won't bother you."

Maybelle: (Tearfully) "I'm not going to give it up if I can help people. I think there's somebody out there that needs me ... that really needs me. And they're asking for help. I can't abandon them, and I can't stand it. But I'm not a person that can ignore a thing. I can't shut it off; I can't turn my back on her."

Andrew: "You must learn to protect yourself. If a starving person knocked on your door, you'd feed them. But you don't even know who or where the woman in the coffin is; if, indeed, there really is such a person."

Maybelle: "You guys are here talking and I just heard it again."

Aubry: "What did you hear?"

Maybelle: "Someone calling my name. May ... May *...Maybelle!* I guess I'm going nuts..."

Andrew: "No ... you're not." (With more conviction than I felt.)

Aubry: "Unh-unh, hon. You're not."

Although I'm not gung ho on soothsaying myself, I offered to put Maybelle in touch with a local sensitive who teaches classes in psychic awareness. She'd have none of it:

Maybelle: "Could this have something to do with my not being Christian? Because him and I have our differences about religion."

Andrew: "Of course, I'm not an expert on the religious aspect of all this, but one theory might be that they're trying to get at him through you: trying to weaken his faith in some way. These two or three things it's done so far may be just the beginning. Later on, this so-called Neptunian, who says she knows all about Christianity, may try to convince Aubry that his faith is all wrong, or all mixed up. I don't particularly buy it; it's just a theory."

Aubry: "Okay. If I get this cross and we still have occurrences, should I call you?"

Andrew: "Oh, you'll still have occurrences. You might lose your space friends using the cross, but they're not going to give up. You're not going to let them. I know that before I leave here. You're deeply rooted in all of this, partly due to the way you interpret your religious beliefs. Also, although I don't want to muddy an already murky stream, you probably have a psychological need for this stuff."

Aubry: "You mean I'm nuts?"

Andrew: "No. Although I don't really have enough information about you to make that judgment. If you're crazy, then a lot of other folks I've talked to were crazy. I don't think they were."

Maybelle: "If you tell Aubry something, like not to do, and it's going to help me, he won't do it because he does love me."

Andrew: "If I could advise him on what not to do, and if I thought he'd listen to me, I'd tell him not to be involved in this anymore; not to read about it, or discuss it with anyone. And, in general, not leave himself open to it anymore."

Aubry: (Missing the point by a mile) "Okay, let me try to figure out what you're telling me. You're saying that I'm always going to be oppressed by evil. Right? Am I going to be oppressed more than the average man?"

Andrew: "I think as soon as reason comes back to you; as soon as you become more normal in your thinking about your religion and about the unlikelihood of Neptunians speaking through your wife; as soon as you are able to place yourself in space and time once again, then you probably will lose these disturbances. Right now, it all fits within the framework of your belief system. But these things are not normal, not even in the context of the paranormal world. I think you know that."

Aubry: "Okay. I see what you're doing. What you're doing for me is giving me the objective side. Okay? Which I'm subjective in this: I can't always see all that's happening around me."

The Wilton's complemented one another almost perfectly. Maybelle supported her husband's testimony up to, but not including, the Neptunians: she may have felt that Morla represented a threat to her marriage. He attested to May's clairvoyance without reservation. Regrettably, I didn't hear from the children. They may have had an entirely different tale to tell.

Was it all an elaborate joke? I didn't think so at the time, perhaps because it was too complicated for that. To my relief, neither of them was the least bit interested in a Psi Session. Inducing a trance-like state in the overly suggestible Wiltons would have been unconscionable. The interview had left me doubting their emotional stability. I was determined not to test its seemingly fragile nature. In any event, *he* wasn't prepared to accept a paranormal explanation, and *she* was totally opposed to a psychological one. Talk about your "belief systems." Aubry, still basking in the "Divine Light," was beyond my help. His wife? I'd seen that sparkle in the eyes of other self-styled psychics. Aubry was convinced the Lord had put it there. For all I know he was right. Mrs. Wilton may have been a female Nostradamus, but she came across as a frustrated housewife whose existence had been enlivened by her newly acquired "gift of prophecy." The late psychic entertainer, David Hoy, used to tell the story about the time he did a TV show with a nationally known "seer" who, it was alleged, was divinely inspired. Responding to her claim that God's spirit worked through her to create visions and prophecies, Dave said "That's marvelous ... that is when you're right. But it sure makes God look bad when you're not."

Before leaving the Wiltons, I suggested that they get professional counseling: "Tell your doctor what you've told me." I hoped he'd know where to send them. They declined. I knew they would. Aubry kept me posted in the days that followed. Maybelle's dream eventually faded away. "The squeaking wheel has been oiled!" she announced, shortly after learning that mortuaries always embalm the dead. Whether on their way to crypt or crematorium, they were beyond Maybelle's help. Aubry never mentioned his Neptunian visitors again; neither did I.

I don't know why all these apparently unrelated events came together the way they did. Thinking back to the few that were corroborated, it looked like a Directed RSPK haunting with Random overtones (at one point, the bedroom window slid open by itself after Maybelle complained of a lack of air). And the Focus: the family collectively, with a combination of individuals providing the motivating force. It's almost as if there's an intelligence somewhere that says, "What can we do to these poor mortals. What will they accept?" When Aubry wouldn't buy demons anymore they converted to Neptunians. You'd think he wouldn't buy Neptunians–but he did. As always, the bottom line is who or what is directing the intelligence?

Manufactured Memories and Unintentional Lies

All this talk of hypnosis smacks of, if not moral then certainly scientific duplicity. We know that an entranced person will, if possible, always tell the truth. If not possible, his subconscious will create "truth": it will manufacture memories to present a suitable story to the hypnotist. During the Psi Session I ask leading questions of those in a frame of mind highly receptive to suggestion. In effect, I direct them to supply answers that appear to satisfy my demands, much the same as the entranced person is commanded to respond to his, or her control--the hypnotist. (I should probably amend this statement to read the "hypnotherapist," since I never ask my clients to bark like a dog.)

Again, the Psi Session is not exactly hypnosis. Yet the atmosphere and ritual of the seance promote hypnotic induction. We know that people can concoct extraordinary scenarios in response to suggestion. Even those in light states are prone to answer questions in an excessively cooperative manner. They may construct elaborate stories of invading spirits to satisfy the desires, or imagined desires of the hypnotist, or in my case the person conducting the ceremony.

Dr. Jacques Vallee, author of the trilogy, *Dimensions, Confrontations,* and *Revelations* (Ballantine Books, New York, 1990-91) has pointed out that anyone conducting experiments in hypnosis with a vested interest in their outcome will have difficulty getting at the truth: "They are wrongly assuming that they are looking at the first level interpretations of what the witnesses have encountered," says Vallee.

Another shortcoming of hypnosis--lying--is addressed by associate professor Howard W. Timm of the Center for the Study of Crime at Southern Illinois University: "Lying under hypnosis is not unusual and not necessarily intentional, either. Subjects tend to fill memory gaps with invented answers." Apparently, their lies are so convincing that even experienced practitioners have a difficult time spotting them. Timm's remarks are important to me and even more so to students of reincarnation who believe in hypnosis-related claims of prior lives as generated during life-regressions (reliving an alleged past life under deep hypnosis).

It goes without saying: If the information received is the result of elaborately constructed inventions--whether deliberate or not--its evidential value is nil. Yet I know of cases in which trauma was discovered in a "prior lifetime," treated by the psychoanalyst as though it were current, with the outcome that the symptoms were, if not totally cured, considerably alleviated. It has never been my contention that the end product of the Psi Session could prove any particular premise except the suggestibility of the human mind. If by treating the information I receive (no matter how fanciful) as though it were factual I can remove the haunting influence or relieve some portion of the discomfort it causes, then it's important that I do so. Since my goal is to help my clients, my responsibility is to them--not to proving or disproving the evidential worth of what they say. I am not particularly concerned with protecting my scientific credentials from the scorn of academia.

Another practice bound to draw criticism concerns those times when I roll up my sleeves and join in the seance. It's a rarity, but sometimes when the night is growing old and the messages are slow in coming, I will join in the writing planchette or table experiment. I'm not a shill. I don't sit in on the game to influence it, only to get the process cranked-up a bit. It only takes a couple of light, *intentional* pushes to "prime" the subconscious mind and open the way for better results to follow. It's another trick of the hypnotist that works. Apparently, one or more of the regular participants take up the suggestive nudge and carry on from there, at which point I bow out.

I'm seldom asked to a house simply to entertain the residents; but it happens. I count on the honesty of my clients. Above all, I rely on the evidential content of the messages received. The proof of a genuine link with a ghostly energy source (whatever that source may be) is in the information it discloses. If evidential, it will contain information unknown at the time to any of those present but verifiable later. It will also include the reason behind the haunting and, hopefully, give me an idea of how to bring it to a quick end.

Chapter 6: Busting the Ghosts: A New Approach

Besides its value in promoting psychic phenomena, the Psi Session is a new approach for an old ritual: Exorcism. It is a modified form of the ancient religious art of spirit expulsion. In modern times, ritualistic exorcism is the last assumption organized religion makes (mental illness being the first). Permission for the ceremony, if it's given at all, comes only after a painstaking investigation. Since exorcists are often criticized for making things worse, the Church's hierarchy rarely agrees to get involved.

For many, a well-structured religious ceremony is the preferred way to chase spirits. Yet, if an invading psychic force can be exorcized, a layman can probably do the job as well as anyone. I don't mean to sound presumptuous, but since a mental house cleaning: getting rid of useless things like repressed frustrations and anxieties can stop most hauntings, the Psi Session can be just what the doctor ordered. After it's over and I've left the scene, each individual tends to become his own personal therapist by making whatever changes are necessary to get his mental house back in order.

The Psi Session permits me to access that region of mind below the threshold of conscious awareness. Participants have little difficulty maintaining an altered state (within Alpha-Theta brain wave levels) long enough for me to uncover information. Success is gauged by the use I make of that information. I banish some alleged haunting entities and Directed RSPK effects by giving the source a chance to be heard. I've had little success with Psychic Imprints or Random RSPK, though there were times when mentally-charged phenomena abruptly ended once the Agent accepted that he or she was the cause of it.

When apparitions and ghosts prove to be images replayed on Nature's recorder and not creations from mind-storage on the loose, it is impossible to reach them through the Psi Session. I can't communicate with what F.W.H. Myers called "veridical afterimages": recreated moments from the past. Nor can I reach Random RSPK, those moronic, purposeless outbursts of psychic energy. When I try what ensues, like the drivel that sometimes comes from the Ouija board, only adds to the confusion.

Earlier I emphasized that those who witness Imprints are likely to do so for as long as they live in the house. Energy patterns have been known to repeat sporadically over centuries and probably account for the ghost-ridden castles and mansions so popular in Gothic literature. Parapsychologists know that investigations and especially attempted exorcisms of Random RSPK (mainly the Poltergeist) have backfired; what in the beginning were only annoyances rapidly deteriorated into bedlam. Since they're caused by a mindless energy I'm unable to make *meaningful* contact.

Even if I could, RSPK seems to provide a safe though exaggerated way to vent anxiety and frustration. These wild demonstrations may be keeping more serious emotional disorders from surfacing. Interrupting the process before it's run its course could be a disservice to the individual and his family. Moreover, since RSPK may indicate the presence of mental dysfunction, my efforts could prove dangerous to the Agent. Having said this I should add that even if I wanted to take these cases, as a rule they're history by the time I get wind of them.

There's another side to the Random issue, one that I touched upon earlier. I don't actively participate in them (above all the adolescent variety) because of the "family culprit" label. Identifying the Agent, usually a bewildered youngster, can be tantamount to putting them on trial. Of course, they're guilty of nothing and, ordinarily, unaware of their role in the fracas. Since it customarily centers on a child, it is difficult to dissuade some parents and siblings from inflicting physical and emotional punishment on them. On more than one occasion I've heard a father or mother say, "I'll beat the holy hell out of that kid if he doesn't stop it." What with the added stress and anxiety of puberty, the very act of naming him may result in even more fireworks.

During outbreaks of Directed RSPK (the reader will recall that these are tame when compared to the Random ones) the Agent is usually an adult. Just the routine of an investigation may interrupt the cycle of phenomena or change the way the observers react to it. Although at times I'm forced to play the "ghost game" with the family before it will turn them loose, left to its own devices Directed RSPK ends up in pretty much the same way as Random: it loses momentum and leaves on its own.

The good news is that neither haunting entity, RSPK, nor Imprint pose any threat to the homes and the people they visit. An explanation of this fact is usually sufficient to return the household to its pre-phenomenon status.

Seances and exorcisms don't always do the job. Not all presences can be chased; and the ones that are susceptible are not always willing to go. Be that as it may, with or without the formal ceremony I leave my clients in a considerably better frame of mind after my visit then they were in before it. To me, that's the true test of my abilities.

The Left Brain Doesn't Know What the Right Brain's Doing

Science is only just beginning to understand how the brain functions. In the 1950s, Roger Sperry and others at Cal Tech did an in depth study of its two hemispheres. Working with patients whose cerebrum had been surgically split Sperry found that most of our awake lives are spent under the control of the predominant left-brain hemisphere. The left hemisphere controls the right side of our physical body and contains much of the verbal, logical, and other conscious mind functions. The right-brain hemisphere controls the left side of the body. It functions in an intuitive way that is clearly subservient to the left.

Sperry believed that the left became the organizer and controller of our daily lives due to a slow process in evolution. In early times man probably didn't need a developed left-brain to survive his environment. Acting more like his primate cousins, he relied on the right hemisphere, on instinct to survive, and on telepathy to communicate his thoughts and fears.

As the centuries passed and he evolved into a mentally complex being, man's right-brain hemisphere slowly wasted away while the left grew to take over its functions as well as a whole group of new ones. The right is thought to create images, and be responsible for intuitive behavior, hunches, dreams, and irrational acts. It is also the suspected seat of the subconscious mind.

I solicit right-brain hemisphere responses during the Psi Session. The ceremony includes techniques of arousal-control which totally absorb the participant in the non-verbal activities, thus shutting off the left-brain's verbal consciousness as completely as possible. It acts as a suppressant to the normally dominant left-brain hemisphere permitting access to the right. If indeed the right-brain is the center of psychic activity--perhaps the seat of the subconscious mind--and if this level of consciousness is aware of the source of the haunting, then success at the seance table should follow.

We're told that the right-brain contains our primitive characteristics. It's more emotional and seems to behave in a way that points to psychic abilities. What's more, if Dr. John Eccles' (State University of New York) theories are correct, in left-brain predominant people (which most of us are) the left hand can write automatically.

In a *Science News* article Dr. Eccles said, "If a design is presented to the person's visual field in such a way that it registers only in the right half of the brain, the left hand will copy it automatically. Unless he watches his left hand the person has no consciousness of what it is doing."Apparently, our right hand doesn't possess this ability, since it operated hesitantly or not at all. Meanwhile, the left, somewhat annoyed, tried to help the right hand along. What does Dr. Eccles' findings mean to the Ghost Detective? After I read the article I introduced a simple test to find the predominant hemisphere for each seance sitter. If he or she was right-handed, I learned there was a 98% chance of their being left-brain predominant. If they were left handed the problem was slightly more complicated.

We've all seen how lefties hold their forearm across the top of the paper and bend their wrist at a sharp downward angle perpendicular to the forearm. The posture of bending the hand 180 degrees around to write is called "inverting," and there's a very good reason for it. The majority of lefties, like most of the rest of us, receive writing instructions from the left-brain hemisphere--the side that controls the right eye. Since the right eye is the "window" to the left hemisphere, lefties are simply positioning their writing hand as close to it as they possibly can.

Probably because it's so rare I've never come across a righty who inverted his hand around toward his left eye. And, as far as I can recall, only once was a left handed participant right hemisphere predominant. How could I tell? He wrote in a non-inverted position, straight up and down. Hemisphere predominance is important because of Dr. Eccles' theories on the left hand's ability to operate automatic writing devices.

When my clients sit down to the seance table it's a good bet they're left hemisphere dominant. According to a 1989 study, 87% of the population are right-handed *(Reuters News Service* article by Professor Jon Ferry). Because the left hand is controlled by the right-brain hemisphere (the window to the subconscious mind) I instruct them to use their left hand to operate the devices. Does it work? I have no statistics to prove it, but it appears to.

Yet I would be less than honest if I failed to add that roughly half of the studies following the 1950s showed no significant differences between the hemispheres. Furthermore, experiments have failed to show that the right hemisphere is the sole seat of psychic activity or of the subconscious mind. In the early days of research they were likened to two distinct persons, each with different talents. Today, scientists generally agree that the functions of the brain should not be divided into compartments. Those previously thought to be right or left regularly appear in the opposite hemisphere, suggesting that both sides share equally in the work load.

The Unsupervised Use of Séance Devices and the Denial of Reality

The improper use of talking boards and other mind-opening devices has led to a number of calls for help from those addicted to them. After making too great an emotional investment in their use, many become convinced that they're possessed. There's nothing inherently evil in Ouija boards and pendulums as long as they're used for entertainment. My advice, as with all useless habits, is to quit them cold! When it dawns on them that they're talking to themselves, most people are quick to come out of their self-imposed stupor and the spell is broken.

Psychology recognizes the part automatic writing and other involuntary physical responses--movements made without conscious volition--play as a method of exploring the mind. At the same time, it knows there's a potential danger when these devices are operated by the inexperienced. Tools that lay bare the subconscious and subject it to the probing of amateurs are not safe. It's the same valid concern that has been directed toward hypnotism for nearly 200 years. Much of the material below the conscious threshold consists of repressed thoughts and emotions. The use of automatisms overrides the mind's ability to censor it. Some contend it's unhealthy for repressed material to be uncovered; that it was originally buried to protect the mind from becoming overwhelmed.

Obviously, any psychic event in the hands of a neurotic or psychotic person is a dangerous one. Even "normal" people are at risk, particularly the young and impressionable. But the danger lies in the mind of the operative, not in the pendulum, planchette, or table; they're just inanimate objects.

I will not knowingly take a case that includes a person undergoing treatment or one who is otherwise mentally unstable. Nevertheless, there is always the chance that a disturbed individual will be included among a group entering the seance room. If during any phase of the investigation someone exhibits pronounced abnormal behavior: anxiety, extreme fear or excitation, unrestrained maudlin and crying, or uncontrolled laughter--it would be sheer folly for me to continue. Only after I've excused this individual will I consider going ahead with the proceedings.

The Psi Session is a modified form of exorcism and a very serious business. The unsupervised use of its devices can lead to a denial of reality and even schizophrenia. An attorney friend of mine once admonished me, saying, “You gotta be off your rocker to play ‘The Exorcist’ and put yourself in such a vulnerable position.” It took a while but I got the message. I still employ it when absolutely necessary, but nowadays most of my cases are solved by means other than century old parlor games and even older religious rituals.

Chapter 7: In the Grip of Evil

"Round and round the circle, completing the charm
So the knot be unknotted, the crossed be uncrossed... And the curse be ended."
(T.S. Elliot)

Case File: Binding the Devil with Fire

At first, no one suspects Evil haunts the portrait even though all who approach it are repelled by hate. In time, uncontrollable anger, fear, revulsion, and even death comes to those under its influence. The plot of Ghostbusters' II? Yes. But also the tragic experiences of the Flynn family in a case I call, "Binding the Devil with Fire."

In the Columbia Pictures' sequel, Prince Vigo "The Unholy," a sixteenth-century Carpathian sorcerer, stands like an ugly Blue Boy in sepia scowling-out at all who dare gaze upon him. From the Metropolitan Museum of Art a river of "psycho-reactive mood slime" flows outward spreading ill will among the inhabitants of New York City. If left unchecked the whole world will be infected with a sort of creeping belligerence.

In my living scenario, the artist is an inmate of the Florida State Penitentiary and his subject--the most improbable source of malevolence imaginable: The Crucifixion of Jesus Christ. You might expect to find the seeds of destruction in the works of Renaissance artists like Bosch, Brueghel and Durer, but hardly in the art of a twentieth-century painter; and certainly not in the most symbolic of all Christian scenes.

In September 1994, I traveled to Plant City, Florida to visit the Flynn family (all the names in this case have been disguised), just about the most amiably pleasant group of folks I've ever met. Earlier that year, Patrick Flynn, Jr. had passed away. He was barely twenty-nine. The pain was still etched on the face of Patrick, Sr., the young man's father.

"It's ironic, isn't it?" he began. "Patty was a Green Beret. He saw plenty of fire-fights in the Gulf War, but he wouldn't stay in *that house.* .. not overnight, he wouldn't. "He came home from the war without a scratch ..... with citations and honors (arms outstretched) this long ... and then to come down with the big C." Grimacing, he continued. "It took some doing, but we've come out of it ... the confusion of grief that's mixed with the business of burying a child. But enough of that. You'll be wanting to see where we lived."

We drove a mile or so to the abandoned two-story frame house, a large, bright looking place built in the late twenties. A white picket fence surrounded the rectangular yard; I could smell a recent coat of paint. It looked to be in move-in condition: a cream puff. So, I had to ask, "Why's it empty?" Patrick nodded, and turning to face his wife said, "It's a shame, isn't it. But you'll have to speak to Beth about that. Personally, I don't see why. " We entered the foyer:

Elizabeth Flynn: "We raised a baseball team plus one here. Every time I come by, all the old memories come flooding back. I hated to leave, but we can't sell it now for the same reason we had to leave. I dunno, by now the evil may be gone. I really don't know! But when we were here I felt something else was here with us-something that shouldn't have been. Why should someone else have to suffer? We can't sell ... not until we're sure it's over! "

Patrick: "We had ten children, altogether. When we moved in '77, let's see ... Patty, the oldest was twenty-seven and wasn't living with us...and the youngest was ten; that would be Michael."

Elizabeth: "I asked for the painting myself. It used to hang over there ... on the shelf above the fireplace. My daughter and son-in-law, the ex-priest, had a convict at the penitentiary do one for us like he did for them. It was quite a while before we finally got it. The first one he did was destroyed in a flood at the prison. The second was lost in a fire. When we did receive it, it was not the crucifixion I had seen from my daughters picture. It was different."

Andrew: "Was it supposed to be the same?"

Elizabeth: "That's right. He was to paint it, you know, the same as he had painted hers."

Andrew: "How is it that the family knew this man?"

Elizabeth: “Before my son-in-law married my daughter he was a Catholic priest, and he would visit the penitentiary and talk with the convicts. He knew this man, this Joseph Penay from his childhood. They were both raised in Alberta, Canada. Joe, who was from the Islands somewhere, was in there for drugs. He tried to escape and he killed an officer, a policeman, and was back in jail for even a longer sentence”. “While in jail he took up the hobby of painting. He did some beautiful work ... work of clowns. When I saw the crucifixion he'd painted for my daughter, it was beautiful and had a lot of color in it. So I asked her if he would paint us one."

Patrick: "When it came we were really startled by it."

Elizabeth: "When we unwrapped it, it kind of took me by surprise because it wasn't the same design that I had wanted. It was in a lot of black. The whole background was black, and the red from the blood area-like from the hands and the crown of thorns--was a bright red."

Andrew: "Would you call it gaudy?"

Patrick: "I sure thought so. But we hung it anyway." We entered the empty living room:

Elizabeth: "We hung it here ... over the fireplace mantel. And that was the end of our peaceful home. After that I never felt comfortable in that room, or in the dining room again. But I could never figure out why.

"The living room and dining room was always our 'fun-time' place with the family. We always gathered there. But shortly after we hung that picture, it seemed like things were beginning to turn around. Dad couldn't walk in the door that we weren't at each others throat. The kids said it was just *terrible.* They thought we was going to kill each other. We had arguments, and they would fight amongst themselves. He and I fought terrible. We had such a fight one time, like they'd never seen us. It was just not like us. Oh, we had our little spats, but nothing like the knock-down fights after the painting came."

Patrick: "She's right. We always got along pretty well before."

Elizabeth: "The room had kind of a cold feeling in it. There was never that warm, homey feeling anymore. As a matter of fact, a couple of the girls ... my daughters said they could just *feel* this chill when they walked in the room. "After I'd get all the kids in bed, I used to be able to sit on the couch and say my prayers at night. But with the picture there...I would look up at it and it would just turn me off. I just couldn't concentrate anymore. "Our bedroom was right behind the fireplace. At night, I would hear noises coming from where the picture hung. I thought it was, you know, like it was a bird that got in there. But I'd never heard any noises before.

"But the main thing was the fighting and arguments and the complete change of character that took place. In the summer of 1986, young Pat came by the house one day, and the kids informed him that mom and dad just fight constantly over things that we would normally have overlooked. The happiness was gone. It's hard to explain. We were a close-knit family and then it was gone. "He never told me then, and it's too late to ask him now, but I think the talk we had that day changed young Pat's feelings about the house for good. I told him that there's something about that picture that gives you the creeps. Not long after, he went to see Father McGee to talk about our problems. Father didn't hesitate one minute... He said, 'You go home and you tell your mother to burn that picture immediately. There's evil coming from it. .. evil forces that are trying to break this family unity up.' It's like a family that prays together stays together, and this evil thing was against that idea. Whether it came from the man who painted it or from somewhere else, you know, no one can tell.

"When young Pat told us what Father said to do, I called my neighbor, April. She was always like an aunt to my children. She's part Indian and part Irish, and kind of like a psychic type person. I told her the story and asked if she would come in and take a good look at the picture. She took a quick look at it and walked out right away. She said, 'I have to go home and get my crucifix.'

"When she came back, she asked all the kids that were home to line up around the fireplace. Let me see ... there were eight of us around the fireplace at the time. And April took the crucifix and put it towards the picture, and as she did it pushed her back with great force. She did it again and two of the boys Michael (nine) and Timothy (twelve) stood and held her up to the mantel; and the force of that picture was just pushing all three of them back.

"So anyway, Timmy took the picture down and put it in the fireplace. April said, 'Hold hands and make a chain and pray, but whatever you do, don't break the chain.' Well, Tim poured lighter fluid on it, but we couldn't get it to burn. It was an oil painting and should have gone up one, two, three. And ...I don't know how much more stuff he poured on there. And then he said, 'Excuse me,' and said, 'You *son-of-a-bitch,* you better burn!' And with that it caught fire. It was like a tornado was going through the house when it did. The doors started opening and slamming shut by themselves. A cold, cold air came through. And as we looked out the living room window, it had turned dark outside ... you know, like we were having a storm. It came on all of a sudden. Later, when I asked my neighbor about it, she said no one outside had seen it get dark. And the doors, you thought they were going to come off their hinges, they were opening and closing so hard. And it burnt. And as it burnt, an image of what each of us thought Satan looked like showed up to the left side of the Lord. As this image was portrayed, you could see like a side view of it...a silhouette. There were horns, pointed ears, and a goatee. The ears reminded me of Doctor (Mr.) Spock, from Star Trek. There was a devilish grin on his face and you could almost hear him laughing.

"But we just kept praying. Some were praying to St. Michael, some to St. Theresa. April kept telling us, 'Don't break the circle.' And the force of the wind in the house was so loud and felt so cold, and the heat coming from the fireplace was so hot and foul smelling. "It seemed like it took a long time for it to incinerate, but it just would not completely burn. Finally, when it was finished, a complete calmness came over the house. The doors stopped slamming; there was no cold air; it was a warmth...a friendly feeling. We were all exhausted. It seemed like it took all our energies out of us. Father McGee told us to bury the ashes, and we did that same day."

Andrew: "What happened in the house after you burned the painting?"

Elizabeth: "Nothing. Absolutely nothing. Except, right after, young Patrick came down with cancer and you know passed away early this year." She winced, and there was more than a hint of uncertainty in her voice when she added, "But I can't blame that on the picture." We stepped back into to the foyer:

Elizabeth: "After the painting was destroyed we had a friend of ours who is a priest come in and bless this place. I had told him about the painting while we still had it. He'd seen it, but I guess he didn't want to get involved in such things before."

Andrew: "Pat, you told me that something else happened ... something that made you both feel this Joe Penay was evil."

Patrick: "Now this part is really hard for me to speak of because I get so angry." Elizabeth: "It's hard for both of us. But this son-in-law of mine, the ex-priest, he got this convict released from prison. Don't ask me how he did it, but they let him out. Penay came to live at their house ... in my *daughter's house!* The day after he got out of prison, after my son-in-law went to work, this man went out and got another ex-con friend of his, brought him to the house ... had the ex-con take the two children and keep them downstairs, and he took my daughter upstairs and he raped her all day long!

Patrick: (His face at last beginning to lose its crimson shade) "This house was really something a few years ago. Young Pat was in the Green Berets. I told you that didn't I? Well, rest his soul, when he was still in the reserves he was injured in a training explosion. When he came out of the hospital I asked him to stay here with us while he recovered. Patty wasn't afraid of anything in this world (repeating what he'd told me before), but he said to me, 'Dad, I would rather not ... I'd sooner go back to Kuwait than stay here.'"

The Flynns are good people, perhaps too good. I got the feeling they'd carried the burden of the old house, especially the financial burden, beyond reason. I tried to convince them that whatever had happened before was gone now. The destruction of the painting- putting its ashes in the ground and getting the house blessed afterwards--should have done the trick. It was time to sell and leave the past to the past... In June 1995, they took my advice: "For the longest time I just couldn't bring myself to unload all that evil on somebody else," Elizabeth said uneasily. What had happened to Joe Penay? "They caught him and put him back in jail, you know, for hurting my daughter. But now, I understand, he's back on the street again. What else is new?"

The Flynns told me they wished I'd been there to work with their *demon* painting. If I had, I would have persuaded them to return it to the artist. Failing that, my advice would have echoed Father McGee's: "Burn and bury the 'sucker' as fast as you can." During the brief period that Joseph Penay's painting hung in the Flynn's living room a contagious hostility hung with it. And yet, even before they centered it above the mantel--when the canvas was delivered and the wrappings removed--Pat and Beth felt a certain antipathy toward it. They were startled they said by the colors: the bright blood reds and the sinister looking blacks. But colors aren't intrinsically evil, or are they?

Perhaps by coincidence, perhaps not, black and red are the two principal colors of voodoo magic. In those places where the ritual is practiced, Catholics and Fundamentalist Christians alike hold voodoo to be the work of Satan. In the fall of

1994, I wondered if they had somehow offended the religiously devout couple even before the painting began to undo the unity of their family.

Voodoo is the black man's version of witchcraft. The practice is known in some of the emerging African nations; in Brazil; England; the U.S., and especially in the "Islands" where, it was said, Penay originated. Practitioners are credited with the ability to cure illness, inflict injury or death, and re-animate the dead who come plodding back from the grave as *zombies*.

When Francois Duvalier took the presidency of Haiti in 1957, one of his first official acts was to redesign the national flag, incorporating the two voodoo colors--black and red. "Papa Doc" was a physician, an educated man and probably didn't believe in voodoo himself. But the peasants did, and he used this strange mixture of Catholicism and African native magic to keep them in line. One method of controlling the peasants was to give his secret police, the dreaded Tonton Macoute, complete authority over the practice of ritual magic.

The word voodoo derives from the Creole "vaudou," which in turn comes from the West African word meaning god, spirit, sacred object, or fetish. Peter Raining, author of *The Anatomy of Witchcraft* (Taplinger, New York) called it a "hodge-podge of pagan ritual and Roman Catholic liturgy." Raining tells us that a common prescription, or *botanicas* (as they are known in Spanish) for voodoo magic involves the use of statues of Christian saints and crucifixion scenes depicted in drawings and paintings. Whether there is a connection between these outwardly appearing symbols of the Christian Faith and the evil painting of the crucifixion we do not know, but the coincidence is alarming!

Who Do Voodoo?

In August 1988, I met native Haitian, Celestina Benice (a pseudonym). For reasons I was never to learn, Celest believed she was under the spell of a powerful voodoo priestess or "Mambo" operating out of Miami. The young woman was being tormented, I was told, by an evil curse-charm created from hair clippings and fingernail parings left at the beauty shop. How did she know? The Mambo was advertising it all over Miami. The ride up the two-man elevator to Ms. Benice's hotel room was an adventure in itself. Halfway to the third floor, as if to thwart my mission, it shuddered, jerked up and down and stopped in its tracks. Thankfully, the car only hesitated for a moment, then lurched upwards again.

Voodoo is supposed to give protection against the evil actions of the *culte des moms* --the cult of the dead. Chief among its gods is the notorious Baron Samedi. An effigy of the dreaded Baron--the lord of the cemetery, master of black magic and eroticism--complete with miniature black coat and top hat stood on the coffee table. Much to my surprise, in the middle of my interview, Celestina eased into trance. She rolled over onto the floor and began writhing to the beat of imaginary voodoo drums; her body twisting like a slightly overweight snake. Again, I am indebted to *The Supernatural* for the following:

> *"A 'white darkness' takes over their body and they act according to a particular pattern always followed by people whose bodies have been taken over. In voodoo belief the possessed worshipper temporarily becomes the god. Unlike other Westerners who generally regard possession as a sign of mental illness or the work of hostile supernatural forces, voodoo followers greet it as a welcome experience--a means of interaction between humans and gods."*

And, borrowing from *The Anatomy of Witchcraft: "At the climax of the ritual she is penetrated by the god; she writhes; her whole body is convulsed ... there is a nervous trembling which apparently cannot be controlled."* (The "penetration" was entirely metaphoric, I assure you.)

Somehow Celestina survived her "curse. She found a Santeria (the Cuban version of voodoo) practitioner who sent her off with seven kidney beans and a Mason jar, instructing her to "place one bean in the jar every day for seven days; then wrap it in aluminum foil and bury it in the backyard."

I guess it worked. At any rate I never heard from her again.

Possessed by Evil

Ever since the ancient Israelites returned from their Babylonian captivity the standardized conception of man's supreme adversary, that which is perpetually set against him and his gods, has been the Devil. I have no philosophical or religious hang-ups about "evil." I'm not a demonologist or a fundamentalist; nor am I prone to give the Devil his due even when I see the traditional signs of diabolical interference. This work would be incomplete, however, if I failed to discuss events that to the uninformed have all the earmarks of being inspired by "man's archenemy."

There are those who believe all ghostly phenomena are inspired by evil. There's not a case in this book, serious or silly, that couldn't be labeled that way by someone. All have the stamp of Satan on them if you look with eyes squinted. Several years ago I worked with a couple in Atlanta who had a harrowing encounter with evil: a brush with something they did not believe in, yet feared nonetheless. It all started when Donny, their seven-year-old, began having *pavor nocturnus:* "night terrors." It's a condition in which hallucinations cause intense agitation and wild behavior in children. While appearing to be awake, Donny would run around the house in a blind panic with the "screaming-meemies." Then he'd stop suddenly and his face, his whole being, would take on the grotesque contortions of a sexual assault victim.

Choking back tears his mother told me, "I get a sickening drop in my stomach every time I think of it. Something none of us can see pushes him over on his hands and knees and rapes him ... sexually rapes him in front of our eyes! He screams and screams and suffers so much. But we're powerless to stop it. Thank God there's no medical evidence of penetration." The almost nightly ordeal was preceded by deep guttural sounds and giggling that sort of bubbled-up within the boy. "But it isn't *Donny,"* she assured me. "It's coming from something inside ... way down inside him."

"Nearly as frightening is the filth coming out. He says things like, 'He f**d me in the butt,' and 'F*** you, you son-of-a-bitch, in a growling, deep-throated voice; then, 'Mom, mom, I'm so scared.'"

Donny's mother is an attorney; his father, a physician. Long before they called me they had ordered neurological and psychiatric evaluations. At first, he was hospitalized to check for organic dysfunction. None was found. Nor was there any sign of nervous system disease. An electroencephalogram turned up a slight temporal lobe abnormality, but nothing that could account for the attacks.

Tourette's Syndrome, a neurological disorder in which among other symptoms the patient involuntarily shouts obscenities--and one of the most misdiagnosed and neglected conditions--was ruled out. I've referred other parents, driven to despair by the uncontrollable profanity of their kids, to neurologists because I suspected Tourette's; but that wasn't Donny's problem. He wasn't "psychoneurotic": he wasn't grunting, barking, clearing his throat excessively, or coughing and spitting. He had no history of spasmodic contractions (tics). Neither did he experience daytime attacks, always present in Tourette's.

There was no medical explanation. His pediatrician and psychiatrist were certain he wasn't the victim of sexual abuse, a despicable crime against children that we hear so much about these days. What did that leave? As his mother put it, "I guess it leaves you!" They had reached their proverbial ropes end when they turned to me. At first, neither parent had the courage to express their fears. They wondered, half-seriously but in total dread of the answer, if Donny was the target of a sadistic spirit. I reminded them that they were educated people living in the twentieth-century. I advised that they continue along orthodox lines until there was an acceptable solution. They agreed and made appointments with a pediatric neurologist and a child psychologist (the latter a friend of Donny's mother from her university days). But still no answers.

Nothing touches us more than the pleas of frantic parents for their distressed child. Now I'm no great supporter of the claims of demonologists. Yet, in an area where the impossible is often commonplace, who among us can say with certainty what can or cannot be? The history of psychical research is full of unorthodox philosophies. With that in mind, I telephoned Ed and Lorraine Warren, a couple of "ghost busters" from Connecticut.

Notwithstanding the fact that they blame diabolical forces for most hauntings, the Warrens seem to have the credentials necessary for driving them out. As of this writing, the pair are co-directors of the New England Society for Psychic Research and live in Monroe, Connecticut. They were instrumental in forming the Foundation for Christian Psychic Research, an organization of clerics and laymen whose sole purpose is to help the psychically oppressed.

In my opinion, the most complete work ever presented on the topic of evil spirits is the biography of this intrepid duo, *The Demonologist* (Gerald Brittle, Prentice-Hall, 1980). Regardless of the origin of the things they investigate, if the accounts of their exploits are true the Warrens have had astonishing success. Even so, because of the far-out publicity they get I'm reluctant to send my clients to them. Example: A photograph in a supermarket tabloid showing Ed standing over a baby's crib--arms outstretched, palms upward--while a "possessed" Cabbage Patch doll floats above him in mid air. I shuddered at the sight.

I refuse to accept the idea that evil spirits lurk in every dark corner. Ed Warren says that in forty years of investigating spirits he's never seen a scientist go into a house and clear it; yet, he and Lorraine have done so many times. It's my guess that if they used the same brand of hyper-suggestion the Warrens do, they'd "clear" houses, too. And yet in this instance the couple has helped Donny. Ed thought he was bothered by low-level possession, and instructed his mother in a few simple prayers. He also thought it might be necessary to bring the boy to Connecticut. An ordained priest of the Episcopal Church would meet with them and invoke God's blessings; or, if needed, do a formal exorcism to remove the possessing entity by force.

When I spoke to her several weeks later she told me that the prayers had worked. The attacks had stopped and Donny was headed in the right direction. "Even his dreams have improved," she offered. "Clearly, there's no need to take him to the Warrens now unless, God forbid, it starts up again." Even though a crew of physicians and technicians could find no bases for it, I believe Donny's problems were generated by psychological processes. Just as the psychologically inspired Poltergeist will in time peter out, Donny was eventually released. Whether it was the suggestive power of prayer or nature's time clock that "dispossessed" his attacker, the child is on his way to recovery. As for his parents. I wonder if they'll ever recover from the experience.

Demonology 101

Those who spend their lives chasing or being chased by evil spirits often adhere to a literal interpretation of the Bible. Demon hunters and fundamentalists tell us that the "beasts" outwardly resemble the dead, those still in the flesh, religious figures, extraterrestrials, folklore creatures, and inhuman monsters--which just about covers all the possibilities. They come in myriad disguises according to the faithful.

Sightings of apparitions and ghosts are themselves a rarity; maleficent entities are even more so. Some say they're made to endure dark images and malformed figures, while an omnipresent dread permeates their home. They complain of fires; the movement of household items as large as refrigerators; of a heavy, foul-smelling "thing" that comes at night, sits on their chest and makes it difficult to breathe. There's the desecration or disappearance of holy relics and other religious symbols. Physical attacks resulting in bites, lacerations, scratches, severe bruises, and bone and tendon sprains. Reports of being pushed from behind while walking down steps; of being tripped; thrown out of bed repeatedly, or catapulted from a seated position. They're slapped on the face; punched; pinched so hard that black and blue marks appear; or harassed sexually: women and children, as well as grown men. Other equally nefarious sights and sounds are detailed for me, too numerous to list. As the evidence mounts, those preyed upon become convinced their "enemy" is evil.

The classic definition of an inhuman entity is that which "is possessed of a negative, diabolical intelligence fixed in a perpetual rage against both man and God." Although they need little prompting, exposure to holy relics, prayers, references to God, Jesus, or other religious notables provokes them into action. Demonologists pay close attention to stories of misshapen or incomplete human as well as animal forms. Haunting entities are commonly fully formed, passive human representations. They're solo performers caught up in a personal dilemma, either wanting to communicate or, more rarely, to be left alone. They manifest any time, day or night, while their inhuman counterparts show up almost exclusively after the sun goes down. Demon hunters have a theory that mortals are never bothered by these abominations unless they encourage them to enter their lives. An unnatural lifestyle: the use of mind-blowing drugs; couples living together out of wedlock; homosexual behavior; the commission of a crime; dabbling in the Occult--each act, in their opinion, a gilt-edged invitation to unholy domination. They divide all hauntings into three distinct parts:

INFESTATION: The opening scene of a normal--if there is such a thing as normal--haunting finds the human ghost doing everything possible to gain attention. While the inhuman spirit or spirits (often multiple entities) try to create paralyzing fear to break the determination and resolve of their victims.

OPPRESSION: The phenomena increase in intensity and become much more personal and severe. At this point, say the experts, the human ghost is out of its element; the inhuman spirit is right at home. These forces often direct an onslaught of physical effects and lay psychological siege to just one person, leaving others in the household alone. Ostensibly, their objective is to cause loss of control by bringing on horrors so terrible that they break the spirit and faith and cause the individual to become dehumanized.

POSSESSION: When the oppressing spirit has virtually enslaved its victim by the skillful manipulation of his will to resist he or she is possessed. We're told that this was its goal from the beginning: to seize the person, body and soul and drive him to insanity, murder, or suicide. Not content with invading the body, the possessing entity is dedicated to its ultimate destruction. (Source: *The Demonologist,* Gerald Brittle, Prentice-Hall, 1980.)

Psychology refuses to consider Infestation or Oppression as anything but imagined or fabricated events, and sees mental disorder at the root of Possession. Mind-curists call such cases: paranoia, hysteria, or schizophrenia because the outward symptoms--stress, anxiety, disorientation, and fantasy--are similar to mental illness. Suspected demonic possession may also be the result of epilepsy, an amnesic period, a state of hyper-agitation, the projection of multiple personalities, mind-altering drugs, or Tourette's Syndrome. One esoteric theory declares they're elementals or nature spirits of earth, air, fire and water: gnomes, fairies, etc. According to magical lore they're not entities but rather thought-forms that have somehow acquired a "life" of their own. Like Psychic Imprints they've been projected by a former resident of the house. Only in this case they're revitalized by an atmosphere of hate or mental aberration present among the current occupants. I can't help believing that we create and project our own demons in the form in which we fear, expect, or even hope to see them. Similar to all thought-forms, once brought into existence mind constructed beings become difficult if not impossible to control. The French occultist, Eliphas Levi, said, "He who affirms the devil creates the devil." It follows that, regardless of their reality, if you believe in demons you put yourself in their evil power.

Chapter 8: The Terror that Comes in the Night

Case File: The Hagging

In July 1988, a psychotherapist referred two clients to me: a forty-nine-year-old widow and her twenty-year-old daughter. Only six months before they'd fled their home in the suburbs, barely escaping a rubbery looking voluptuously built female with the face of an old *hag.* In the vernacular of the demonologist, what frightened them away was a "Succubus": a wicked, sexual oppressor of men.

Late in 1987, John Brice succumbed to cancer. That would have been horror enough for his wife Lou Ann, and daughter Mary (all fictitious names), but there was more. Sometime during the previous summer, just after his doctor had pronounced his condition incurable, this abomination with boobs began to appear in the Brice home. Lou Ann supplied the details. "When John said *she* was on him, I thought he was dreaming," she said in a thick voice. "The man just found out he was dying; he was entitled to the 'granddaddy' nightmare of them all. Night after night he would wake me thrashing around and screaming, 'Get her off me! Get her off!' Once I saw this glowing outline glide across the bedroom floor. It moved toward him and got up on his chest. I screamed! It looked over at me, hissed, and flew off."

What followed was a series of nightmarish episodes more devastating than death itself. As for the cancer, John Brice had gone through a period of denial before grudgingly accepting his fate; then became as reconciled to it as any middle aged man could be. But the *thing* that kept persecuting him-that was too much! "'She's come to take me to *hell,'* he'd say, but she won't ... not until her appetite is satisfied.'" Whether that appetite was sexual or purely morbid was never entirely clear. "'That's just crazy, John, I'd tell him, 'I can't, I won't accept that.'" "I'm not a religious woman, but I wasn't ready to risk his immortal soul to that *thing."* So a vigil began. The two of them took turns watching over him. He slept alone, guarded round the clock by wife and daughter. As long as they were there to ward off the "hag's" advances, it couldn't take him--not sexually or to hell. But all the loving surveillance in the world couldn't keep John's fragile health from deteriorating.

"He was failing. It was a terrible time for us," she said, choking back the tears. "His doctor wanted him in the hospital ..... and maybe he'd have been better off there ... but he wouldn't go. And I wasn't about to make him. There was no one to turn to. Both our families were in L.A. We hadn't made any close friends since we moved to Florida in 1986. Besides, who would have believed us?"

Then mercifully, the day after Christmas, 1987, John Brice passed away. As long as she and Mary maintained their watch, the hag had kept its distance. They'd managed to keep it at bay for his few remaining days. Now they were powerless to stop it from hovering over his lifeless body. "I had mixed feelings. He was my life. But I couldn't wait for the mortuary to pick up what was left of him!" she confessed. Now it was summer, 1988. Mother and daughter were coming back to get the house in order. Lou Ann called because there was no one to stay with them while they got their belongings together; neither would venture there alone. After she shared the following, I wasn't all that keen on "venturing" either.

"My daughter's doctor saw your name in a directory of organizations. We left in such a hurry...all we took were the clothes on our backs. I was sleepwalking after the funeral: just going through the paces. John's death was so god-awful that, at first, I didn't pay any attention to the things going on around us. Looking back, I should have grabbed my daughter and ran after the funeral...but I didn't. I guess I was in shock." "What kind of things?" I asked. "Well, it was mostly sounds, originally, but then it was more. We muddled through a few more weeks, till the middle of January, but then ... my heart is in my throat just thinking about it..." I interrupted to suggest that she write it out and send it to me if the subject was too difficult to discuss over the phone. "No!" she said in a husky, two martini voice, "If you're going in there you have a right to know what you're getting into. I guess you go through this sort of thing all the time...l mean, in your line of work you see all kinds of horrible things. I really hope so, or you won't believe what I'm about to tell you. You probably won't go in there if you do believe it!

"After the funeral, exactly three weeks to the day, towards evening, I was looking out my kitchen window ... it faces the back yard, when all of a sudden I saw it. It was just standing there, straight up, motionless ... kind of hanging there in the air just above the ground. Its face was ashen, you know really pale, but I could make out its features. (Neither of us ever really saw it before, not without it being blurry.) It was staring at me, looking inside my soul. And when I looked back, I got dizzy; I guess I passed out. When I came to ... it was dark." "Tell me what you saw, if you can."

"I can still see it. It was covered in a loosely draped gown of some sort ... a dull white, and I could see the outline of a well-shaped body underneath. She had gorgeous blond hair, but her face ... it was a two-bagger: it needed two bags to cover it. Ugly can't describe her. When I first glanced-out the birds were chirping; I could hear traffic sounds on the street two blocks away. As soon as it looked back, everything went dead--including me. "Pretty chilling stuff," I offered "But wait! This next part you're not going to believe. I lived through it and I can hardly believe it. That night, JOHN CAME BACK! He was looking for Mary, and when he found her he tried to seduce her. God! I can't believe I'm telling you this."

Now my heart was climbing the rungs of my throat. "You actually saw him?" "No...but Mary did. He woke her fondling her breasts; tried to have intercourse with his own daughter." "Surely she was dreaming," I said in disbelief. "Maybe. But I took no chances... We left for my sister's place the next morning, and haven't been back since." All the passengers were off the plane and headed for their baggage, when a woman in a brown summer tweed tapped me on the shoulder:

"Dr. Nichols? I'm Lou Ann Brice, and this is Mary. I'm sorry, at the last minute I decided not to wear the canary." "But how did you know *me?"* I asked. "Easy. You're the last person left outside the gate. I was half expecting you'd be in a Ghostbusters' jump suit." I took them to the Tampa Airport Marriott and waited while they checked in. Since there was plenty of daylight left, Lou Ann decided to put a few hours in on the house. It was a 20 minute drive from the hotel. On the way we chatted: Lou Ann: "I dread this ...I absolutely do! Going back is not going to be easy for either of us, but it is really hard on her ..... considering what she's been through. (Turning toward her daughter) She is emotionally deprived by what happened."

Mary:"My doctor..."

Lou Ann: "Been in therapy since it began ... here, and in L.A.."

Mary: "I've been..." Lou Ann: "Actually, she's making very good progress."

I quickly formed an opinion of Miss Brice. She was twenty, but she could have been twelve the way she dressed and the way her mother dominated her. Mary was in one of those outfits parochial school girls wear: blue-gray pinafore on a white blouse with sturdy black or dark blue shoes. It turned out that she was quite bright, but a young lady with major hangups:

Andrew: "I'm interested in how you feel about going back, Mary. Is it all right to talk about it?"

Lou Ann: "I don't see why it shouldn't be. Go ahead, dear."

Mary: "I'm interested in what you think about it Dr. Nichols. I haven't told anyone what happened, except my psychotherapist. Of course, he doesn't believe a word of it."

Andrew: "I believe the story your mother told me, Mary ... at least I believe that the two of you believe it."

Mary: "What kind of world is this? I'm sorry, but I'm not stupid or crazy. I know what I saw and what happened to me. My doctor acts like he accepts it. But it's an act; he humors me. How is it that you say you believe, yet my doctor doesn't?"

Andrew: "It has to do with the nature of these things. Some people say ghosts are the dead returned to earth. Others are sure they're sent by the Devil. Most people think they're figments of the imagination. Your doctor is certain they're imagination, or even signs of illness. I think they could be any of these, but I do believe they're real ... real in the sense that the things they do can affect us."

Mary: "Excuse me, but that sounds like more mumbo-jumbo."

Andrew: "Okay. Are you a religious person?"

Mary: "I guess you could say I was."

Andrew: "You know what miracles are ... religious miracles like Fatima and Lourdes? You could call them supernatural. Manmade miracles: men on the moon, heart transplants, cures for diseases--they're paranormal. At one time the manmade kind were thought to be impossible, but they weren't. Ghosts, like miracles, could be either. They could be supernatural or paranormal depending upon what sent them. What makes me different from your psychiatrist is that I'm not sure what sent them; he is. He's positive they're not miracles: that they're just hallucinations."

We arrived at noon. Waiting in their panel truck were two women from Swifty-Sweep, a domestic cleaning service. The house didn't look haunted. Six months of dust and a hundred dead bugs were piled up, but no gloom or doom greeted us. The afternoon was spookless. In daylight, with all the noise and all the dirt flying I doubted that anything would happen. Riding shotgun for them was going to be easy. Packing everything in sight, mother and daughter barely stayed ahead of the cleaning ladies. I taxied them back to the hotel before dark. When we got there, Lou Ann said, "Andrew, you know, I don't think we're going to have any problems finishing this off tomorrow. I don't think we need to waste your time. I'll get a rent-a-car. Besides, the two 'swifties' will be back. They look like they can take care of themselves." I didn't argue with her.

There was no word from them until 11:30 the next night: "It happened!" she blurted. "I can't tell you what happened, but I can tell you that Mary is back where she was in January. It's over! We finished most of it tonight ... and screw the rest. I'm sorry, but I'm through fighting it!" "Tonight! Were you in the house at night?" I asked, not believing my ears. "I'll talk to you tomorrow. We're exhausted. We're okay, just tired and plenty damned scared." I met them in the Marriott Café the following morning:

Andrew: "How is it you were there after dark?"

Lou Ann: "We stayed ...I know, I know, it was a stupid thing to do, but it was getting late and we still had a couple of hours of packing left. And I thought ... since nothing had happened, you know. I asked the cleaning people to stick around, but they couldn't. I sent *her* out on the porch, but she wouldn't stay there alone. I didn't want to have to come back again. We were so close ... so I stayed

"It was something like a quarter after eight ... almost dark. We noticed it got freezing cold ... it was humid outside. The electric was on, but the gas wasn't, so I couldn't put the heat on or light the stove; that would have been out of the question anyway in August. I heard a *wooosh* and looked up. A dark shadow was coming together near the ceiling in the dining room. It became this big ball then swooped down on her. (She took Mary's hand.) I stood there with my mouth open. I tried to get to her, but couldn't move. While I watched it came down and surrounded her. I could still *see* Mary, but the shadow was all over her. Then she said, 'Daddy. NO!' It lasted. .. Oh, maybe a minute or two, that's all."

Andrew: "Is that the way you saw it?"

Mary: "No. Not really. I didn't *see* anything, only I couldn't get enough air to breathe. There wasn't any air. I came in from outside...and it was cold. Then I felt a sharp pain in my shoulder."

Lou Ann: "You didn't tell me anything about that."

Mary: "It's my right shoulder. I couldn't move. I thought I was paralyzed. When I did finally ... look." (She leaned toward me and stretched the neck of her blouse. I could make out two curved rows of teeth marks near her collarbone--eight or ten impressions. The only thing I was certain of was that she hadn't bitten herself.)

Lou Ann: (Pressing forward across the table) "Oh, dear!"

Mary: "It's all right, mother. It doesn't hurt."

Andrew: "What about calling to your father?"

Mary: "No, I didn't do that ... or I don't remember doing it."

Lou Ann: "This attack'll put her back six months. The first one put her in therapy."

Mary: "Mother, I was in psychotherapy *before* daddy died."

Lou Ann: "Mare..."

Mary: "No, I want to talk about it. In most ways, Dr. Nichols, my father was a good man. But he could be destructive, too. He destroyed my childhood; took everything that was beautiful away from me when I was ten-years-old. If my father has become an evil thing it is because he did evil things when he was alive. I loved my dad, but what he did to me when I was little ... what he continued to do to me until I was old enough to stop him ... that was wrong. It took me a long time to understand that it was his wrong and not mine."

Lou Ann forced herself to make one last trip to the house; she had to supervise the movers. On Friday, August 19, I came to the airport to say good-bye. Out of earshot of Mary, Lou Ann said, "I want you to know that I don't think John ever touched my daughter while he was alive. Her doctor doesn't think so either. And I've come to believe that whatever it was that attacked her after he died ... that couldn't have been him either." Although I didn't pursue it, Lou Ann's remark about Mary's doctor (that he doesn't think John molested his daughter) was difficult to follow. Not his opinion, but rather Lou Ann's knowledge of it. Psychiatrists are notoriously stingy with information, unwilling to divulge it even to their patient's parents. But if he's a "traditional" psychotherapist, he is more than likely a Freudian. And if a Freudian, he may have discounted the story of incest. Believing that it was in his patient's best interest, he may have confronted Lou Ann with the accusation--to confirm that it was baseless--then assured her that it was in all likelihood a fantasy.

Early in his career, Freud believed that behind hysterical symptoms lay childhood sexual abuse repressed from consciousness. He consistently dredged up trauma that included incestuous behavior by parents, especially fathers of young girls. Later, he theorized that the symptoms displayed by some women--those who claimed they'd been seduced by their fathers--were nothing more than "wish fantasies": longings that had been repressed in youth, only to awaken at puberty and surface into consciousness in the classic form of neuroses. Today, a few of Freud's detractors charge that he deliberately hid the truth from the public because he was reluctant to face the reality of sexual abuse of children by their parents. He also feared the public outcry to such a shocking revelation if he continued to make it. (Source: *The Assault on Truth: Freud's Suppression of the Seduction Theory,* Jeffrey M. Masson, Penguin Books, New York, 1985.)

There is nothing in orthodox science that can explain the Old Hag's repeated attacks, save illness: physical or mental. Whether the sexual assaults on Mary Brice by her father were an aberration (a wish fantasy) or not, the fact that she *believed* he'd offended her could have set off the deathbed visits of his own tormentor and the frightening events that continued after he died.

THE HAG

Legend says that ghouls, like vultures, are supposed to prey on the dead, not the dying. The fiend that forced itself upon the Brice family and the obscenities that followed John Brice's death was beyond the norm for haunting violence. But if you think the ordeal was unparalleled you haven't heard of Dr. David Hufford. The Old Hag is "old hat" to the good doctor. In his study, *The Terror That Comes In The Night* (University of Pennsylvania Press, Philadelphia, 1982), Hufford examines the evidence for the Succubus and its masculine counterpart, the Incubus, from the scientific as well as folkloric viewpoint. As associate professor of behavioral science at The Pennsylvania State College of Medicine, Hufford is well qualified to pursue this nocturnal troublemaker, a pioneer in the development of a systematic methodology for the study of paranormal evidence. Equally important, he has spent as much or more time in the field as he has in the laboratory researching these widely observed curiosities.

Hufford chose the term "Old Hag" because that's what they're called in Newfoundland where he taught at St. John's Memorial University. Haggings are fairly numerous in that part of Canada and, strangely enough, match exactly a personal encounter he'd had himself. Before publishing his findings he spent the next ten years in Kentucky, Nebraska, California, and Pennsylvania, talking to eyewitnesses to these frozen-faced body jumpers. He approached the topic from an interdisciplinary angle, trying to put together materials from folklore, anthropology, psychology, neurophysiology, medicine and sleep research. The pattern of night terrors among "Newfies" was fascinating because they so clearly went beyond culture and prior expectation. Many victims had never heard of the Old Hag, yet they underwent typical hagging attacks: "It is this feature more than any other, I would say, that runs counter to practically everything that social scientists in this century have said about supernatural beliefs (namely, that naive people see and hear what they expect to see and hear)," he wrote. Folklorists and parapsychologists alike know that academicians routinely dismiss hauntings as being merely the result of expectation and wish-fulfillment. Not so in the case of the Old Hag, says Hufford. He's collected enough contradicting data to allow him to say, "The burden of proof now, based simply on current evidence, lies with those who would like to simply dismiss the reality claims of these and similar traditional beliefs."

Let's take a closer look at what the target of the Old Hag can expect from his "two-bagger" visitor. The scenario goes something like this: The wretched fellow (it's usually a male), who invariably sleeps on his back, awakes in the middle of the night to a crushing paralysis: a "presence" sits on his chest; or, less often but far more threatening, bony hands clutch at his throat. A growing conviction sweeps over him that, unless the unbearable weight is lifted or the fingers squeezing-off his future are removed, he'll surely die. Fortunately, unless he's literally scared to death the nightmarish Hag is no more a threat to life than is the Poltergeist.

Unlike John Brice, when most victims try to cry-out they find their vocal cords frozen, unable to utter a sound. Shuffling noises and deep breathing sounds are often heard in the room. Some tell of footsteps getting closer and closer right up to the moment they're pounced upon. It's a rarity for an attacker to be seen, but when it is descriptions range from glowing light, a gnome with piercing eyes, to haggard looking woman. Sometimes these body pressers transmute between human and animal form. It's not clear if their goal is ever sex. It may be harassment on a grand scale, with sex thrown in to damage their victim's self-esteem.

In the Medieval times, Anglo-Saxons had a term for these midnight misanthropes. They were called *nicht mara* or nightmares. To the knights of old, the words signified "night crushers," not crazy dreams. Although Dr. Hufford began the original study on Newfoundlanders, it was soon clear that the phenomenon was worldwide affecting roughly 15% of the population. Physicians contend that the symptoms are connected with heart, asthma, or emphysema attack, or to any condition that restricts the individual's ability to breathe. Hufford found no unusual numbers of heart or respiratory sufferers among his interviewees, which seems to discredit the theory. Hagging has been improperly associated with a variety of different illnesses (epilepsy for one) but is easily diagnosed once the symptoms are brought to light. Uncovering them is the hard part. Hufford tells us, "Academics explain supernatural beliefs in terms that are very unflattering--so people who are worried about their respectability don't talk about their experiences. When I've spoken (lectured) about the Old Hag, I've had a tremendous response. People want to tell me about their experiences because I don't say they're crazy. They can't be crazy. There are too many of them."

Consistent with folklore, the Old Hag is a supernatural being. As expected, psychologists regard it in much the same way they regard diabolic possession: as symptoms of suggestibility or pathology: "Early twentieth-century researchers, influenced by Freud, ascribed the Old Hag, as well as a host of other experiences to sexual repression and superstition," says Dr. Hufford. As for his own personal convictions, he steadfastly refuses to speculate on what night terrors might or might not be. He maintains, "The contents of this experience cannot be satisfactorily explained on the basis of current knowledge." That's a pretty good definition of what I call the *paranormal.*

More Nocturnal Assaults

Whatever its origin, the pre-dawn terror known as the Old Hag is a menacing force preying upon folks when they're most vulnerable. Added to my list of those who have been hagged. is a high school teacher and amateur historian named James Fisher (a pseudonym). He called me from Waycross, Georgia where, in 1977, he and his wife Beverly had gone to live. I remember the first words we exchanged, how impressed I was by his dynamic personality. I could tell he was sizing me up over the telephone: listening to what I had to say before deciding if I was legit; if it was safe to commit his secrets to me. I must have passed muster because a week later I received the first of his letters. He'd come across an article about me in the October 14, 1994, edition of the *Atlanta Constitution.* It prompted the telephone call, which, in turn, resulted in the letter that followed:

"When I read about your paranormal research project ...I decided to sit right down and write a letter asking your opinion of the nature and possible motivations of 'hants' I have experienced for about five years. I'm not requesting that you come up here and work on the situation--simply give me your opinion, even if brief."

"In the summer of 1989 when the first unexplainable nocturnal visit or attack occurred, I felt that the existence of such things was dubious. It certainly had never occurred to me that I would ever be the target of such forces. So when the first one occurred, it knocked a hole in my armor-plate of skepticism and self-assurance. It shook me up to realize that my psychic immune system was vulnerable enough to suffer an invasion by a force or forces that seemed to be of another plane--invisible and unmistakably malign".

"The first visitation (also the strongest and most unnerving) and subsequent ones (about fifteen in all) occurred around 2:15 a.m. I was awakened by a soundless vibration like a motor hovering over my feet and ankles. It gave off strong vibrations that struck against me. It was manifested as an invisible field of powerful energy that was first apparent at my feet, but soon moved over my person, dipping into some part of my anatomy. As it went it moved in a few seconds up my thighs and stopped over my pelvic region. It then increased in power and commenced to hit against me and rebound in rapid motion, as though seeking to penetrate my body at the genital region. It conveyed to me an ominous and disturbing threat--as though it were an evil force determined to bore into my flesh.

"Until then, I had watched in frozen fascination, trying to yell out, but seemingly unable to speak. With great effort I sat up and swept the covers violently away. At the same time, giving a sort of cry of protest, I made a half-strangled AHHHH RRRRRR exclamation. Of course, it was very disturbing, and I was somewhat frightened. At the sound of my voice and the throwing up of the covers, the field of energy vanished.

"The force appeared again several times, but always moving rather lightly over my person, and then suddenly manifested as a soft object about the dimension of a large house cat (later he told me it was the size a 25 or 30 pound Tom) that pressed against my right side at the hip. As long as I made no sound, or didn't move, the force would stay put. But I could only stand inaction for about two minutes and then wanted to shoo it off. I finally got so I could say 'go away' and 'what do you want?' "Sometimes it would awaken me as a sinister presence in the bedroom. First, from across the room and approaching gradually until it reached the bed. Then it would alight and snuggle down along the outer side of a thigh. "A most disturbing habit has been that of jerking the covers. Once, the covers were pulled a few feet downward." In July 1984, James was paid a visit by a "thing that grasped my feet around my ankles, and gave a jerk. It must have moved my body two to three inches." He began to keep a 15 watt light bulb burning all night at his pillow-side, thinking that the "thing" might shun light: "This went on fine until about ten days ago. I was awakened by something pressing up the two soft pads I sleep on. That is, it was between the mattress and these pads. It moved in slow upheavals about four or five inches at a time--all the way up my body, actually forcing the point of bodily contact to push me upward. It stopped without my saying anything."

After that incident he kept the light on every night for a week. A few days ago, "At about 2:15 a.m., by the bedside clock, I felt a pushing under my shoulder blades only, and the strong force hovering around my feet. It just kept jiggling and pushing against my back in that place for about three minutes. (Even) without the 15 watt bulb (the) night sky light makes the room light enough for me to see any apparition, no matter how flimsy. But so far I've seen nothing. This last time, I strained to see something and also glanced all over the room, but saw nothing.

"I realized I was bracing myself for the visits every night, although they continued to be months apart. I had become angry at the invasions and wanted them stopped." James is a fighter. He had no qualms about confronting "slumber spirits" head on. As he put it, "My intention is to rid the house of the least of whatever it is that is disrupting my sleep." Aware that he needed help to do so he let his fingers do the walking through the yellow pages of psychical research. Inspired by the newspaper article he contacted me. In the months that followed, James exchanged letters with me. He asked if I thought a spirit force had attached itself to his person. I answered:

"The subjective experiences you've been having, without the physical appearance of some type of entity, are not all that unusual. A ghost, if indeed you are being visited by such a phenomenon, prefers to make its debut in a progression of steps. We call the phase you're going through now, 'Infestation.' This may sound like a job for 'Raid,' but seriously, it is not inconsistent with our case histories and not all that uncommon. "As to its progression, a good deal depends upon the concern and attention you pay it. We know that where such psychic effects are ignored (not always an easy thing to do) the chances of a quick termination are measurably enhanced."

In other words I told him to forget it. I approached the problem in terms of a conventional haunting precipitated by psychological needs, rather than a traditional hagging. At bottom, James saw it along these lines, too. His creative imagination, however, took flight when he wrote to me the first time: "I lean strongly toward the belief that these nocturnal visitations are from either spirits, demonic entities, creatures who inhabit regions beneath ice-caps or the seas, or, probing space people." In recent years James has recanted most of these statements. No longer does he believe he was visited by inhuman spirits, and his position on the human variety has waned as well. As to the extraterrestrials, we've yet to pursue that subject. In a telephone conversation he philosophized: "I question that we deserve a Hereafter. Perhaps there is only dissolution and you're only a pinch of elements that melt into the earth. Still, my thoughts are haunted by the intimations of immortality in the words of the great agnostic Robert Ingersoll: "'...From the voiceless lips of the unreplying dead there comes no word; but in the night of death Hope sees a Star and listening Love can hear the rustle of a wing.'"

When it's not considered heart attack, emphysema, mental illness, nervous disorder, or illusion--the "terror that comes in the night" is variously attributed to a demon, a witch, an astral visitor, or even *dyspepsia.* Surprisingly, this last possibility, indigestion, is sometimes the solution to the problem. Until the publication of *Hallucinatory Mullet Poisoning,* an article written by A.H. Banner and Philip Helfrich *(Journal of Tropical Medicine and Hygiene, 1960)* it had been thought that plants and plant derivatives were the only natural source of hallucinogenic drugs, e.g., the peyote cactus; the sacred mushroom; ergot, a fungus, and others. Of course, pharmacology has successfully synthesized hallucinogens for many years, but the natural source of these drugs was thought to be the plant kingdom exclusively.

Now, thanks to the research work of Banner, Helfrich and others--men like J. L. Smith *(Sea Fishes of South Africa)*--we know that some tropical and subtropical fishes produce these effects, too. Apparently, it's a problem affecting various species, but the prime suspects in these poisonings, called "ichthyoallyeinotoxism," are the reef fishes caught in the Carribean. (In fact, more of these reports originate in this region than anywhere else in the world). The specific culprits--labeled "dream fish" by the natives--are two varieties of mullet and two of surmullet, or goatfish, all popular eating fishes in the Islands. Fortunately, toxicity is seasonal, and even in season not everyone who eats these delicacies is poisoned.

What do dream fish have to do with the Old Hag? Experts tell us that when tainted fish is consumed the toxins affect the central nervous system resulting in, among other things, hallucinations, lack of motor coordination, muscular weakness, and partial paralysis. When attacks come at night, the victim can expect to experience a vivid impression of being crushed by something sitting on his chest. It becomes exceedingly difficult to breathe, and the thought that his life is about to be snuffed-out explodes in his mind.

Now here, I felt, was a reasonable answer for a number of otherwise unreasonable events: marine biology versus superstition. Scientific solutions are always preferable to supernatural ones, and I congratulated myself on finding one. Everyone seemed satisfied, everyone except James Fisher. I thought "*Surely someone in the Islands* is *shipping* a *crate of mullet to James from time to time. He's gorging himself on these little delicacies* (J. L. Smith called them 'some of the best eating fish in the world') *and he's paying an awful price."* But James said no, and we were back to square one.

During the period, *1994-95,* James' search for truth led him to an Afro-Carribean exorcist: "Suffice it to say that I had an exorcist in South Carolina lay the main demon ..... via telephone. He is one of the most powerful practitioners of "Root Medicine" in the southern U.S.." In this particular instance, long-distance exorcism turned out to be beneficial: "Visitations continue, but they are fleeting and light of touch. The Root Medicine practitioner explains these visitations as coming from minor, harmless and irresponsible spirits left by the main haunt. And that I am to pay them no attention." "Because of my archeological digs of native American graves, initially a curse occurred to me, but I had every confidence that no curse had ever touched me, or ever would. You convinced me that this thing was not a personalized thing that came from my archeological excavations and so forth; that I had (not) aroused old guardians of a grave and that old curses were (not) working on me ...

because I have the identical common experience that one out of five persons have, but rarely admit. "It may be something going on in my subliminal mind, in my autonomic nervous system ... way down deep in some layer of the Id. It's probably somewhere in the dream mechanism, and it's so real that you have physical sensations of it. However, I would swear that I'm awake when the entity comes into the room. Never have I seen it. There's a sense of being menaced by some evil force that I'm aware of across the room and it comes toward me..." "You're not still having these experiences, are you James?" "No. Oh, no! I'm analyzing it, you see. I have not had an attack ... the nearest I had was about two or three years ago."

Sometime around April, or May 1990, a strong wind had swept across him while he lay in his bed: "As usual, it was around two or two-thirty in the morning. Usually I sleep in the fetal position or on my tummy, because I don't want to go through another one of those visitations. But I had shifted over on to my back. As I awoke I became aware of a wind that was pushing me. I had the feeling it was coming from the closet. It grew gradually into a force that I could pretty well judge to be 40 to 45 miles per hour. Then, at about the time I was thinking that it was going to push me against the wall, it slowly died away. It was a continuous process lasting three or four minutes. "Then six months after this experience, I awakened with a sense of a body on me ... right around my chest and solar plexus. It gradually grew over a minute or minute and a half into a tremendous pressure. I thought of it as a creature that had come upon me. Gradually it lightened and lightened and faded out, like it floated away from me. I had an electrocardiogram afterward (on the advice of his physician) but there was no sign of any kind of problem."

Although most of my clients complain of late night disruptions, I hear far fewer Old Hag accounts than the 15% rate reported by David Hufford (perhaps because many victims do not equate hagging with visitors from the spirit world). All the same, I've had my share. In February 1998, a thirty-eight-year-old housewife told me about awakening to shuffling sounds, followed by the sight of an old woman wearing a long flowing nightgown. A little puzzled by her own description, she said: "Her face was old, but she looked younger ... with blond hair down to her waist. She was coming at me, and the next thing I knew it felt like the ceiling was caving in. She was on top of me, but all I could see were two shiny red eyes. I thought, This is it! *I'm gonna die!* The first time (when I spoke to her she'd already been hagged four times) my six-year-old saw her. When she left, she just floated out the window ... and my kid saw her do it." An earlier complaint was similar, except that the invading spirit sat on the feet of this object of passion rather than on his chest. Still another account was received--this one from New Mexico, in April 1991. Three family members were being hagged.

As for the Fisher case, square one is where the problem stands today. I have no clue as to why his supine body (or anyone else's for that matter) should be periodically worked-over that way. None of the other "experts" consulted were equal to the puzzle either. There's probably enough symbolism in all of these nocturnal experiences to rouse the ghost of "Sigmund." Anxiety dreams, conversion reactions, or some other diagnosis might be cited by the bearded apparition. And who knows, maybe he'd be right. Thankfully, James Fisher is in excellent physical and mental health. As far as I know, he hasn't been hagged since 1997, so perhaps that segment of his life is over. I hope so.

Chapter 9: Body Snatchers

Literature abounds with representations of the supernatural and paranormal. Two of the best are Brain Stoker's, *Dracula* and Mary Wollstonecraft Shelley's, *Frankenstein.* The bloodsucking vampire is an example of *supernatural* evil. Count Dracula was the dread lord of the undead--soul-dead for centuries. According to the legend, this Gothic Batman was subject to oblivion but only under very special circumstances. In the movie versions, the Count transmutes at will between man and vampire bat. Unless you're a madcap romantic, you know that nothing like this re-animated corpse has ever existed in Nature.

Dr. Frankenstein's soulless creation, on the other hand, is a *paranormal* monster. It could exist. Not in the author's time, or even now--yet there is reason to believe that such a creature could exit. It's a repulsive though not inconceivable thought that science will one day assemble an amalgamation of body parts and shock it to life. Shelley's hulk, whose imperfect brain was trapped in a cross-stitched body, was all too mortal.

Two of the great old horror movies of yesteryear were *The Body Snatcher,* based on a work by Robert Lewis Stevenson, and, of course, Shelley's *Frankenstein* (the 1931 version: way before my time but still popular when I was a kid). By today's standards, epitomized by the "slasher" flicks, these two Boris Karloff films are tame; but in their day, they were the stuff from which *nicht mara* are born.

Grave robbery, the macabre theme running through both, used to be reserved for ghouls and mad scientists on the big screen. A few years ago I discovered that you don't have to defile the tombs of the ancients to incur the wrath of the pharaohs. Ordinary, run of-the-mill grave robbers are quite capable of it, even when they've disturbed the sanctity of some run-of-the-mill stiff.

In the winter of 1994, I journeyed to a 145 year old farmhouse in rural Florida. It was one of those times when I found myself in "Kooksville." The beneficiaries of my help were a forty-three-year-old ex-farmer, his wife, thirty-nine, and their four teenage children ranging in age from twelve to seventeen. I'll call them the Bentons, though that's not their name. I arrived on a Sunday afternoon just after two o'clock. The place, overrun with screaming Benton kids and snarling Benton dogs, reminded me of *Ma and Pa Kettle On The Farm.* The racket was deafening, above all, the sorrowful YELPING. Later, I decided that the canine cries were the outcome of a sadistic game the children were playing. They were chasing the little critters around in circles and stomping on their paws whenever the opportunity presented itself. It was pure bedlam!

The farmhouse reflected its occupants. Confusion reigned, but there was something beyond the din. Oddball incidents and oppressive feelings had unnerved the family. First Mark, age fourteen, felt a hand squeeze his shoulder. Then Linda (the seventeen-year-old) began seeing a white misty form--a "puff of smoke," she called it--floating from room to room. She also saw a "white cross" emblazoned on the living room wall. Next, Mrs. Benton (Sylvia) detected the aroma of lilacs in her kitchen. "I smell it a lot of times ... in the early morning. It's a musty, sweet smell like an old lady ... like my grandma used to smell," she told us. "And when it happens, sometimes the kitchen table starts to vibrate." "We used to hear someone crying all the time," she went on, "but no one was there. And the toilet seat--if I leave it up you hear it fall down. If it's down, you hear it come up. And there's the time I lost my balance on the steps. A hand came up and kept me from falling."

They all complained of being stared at by someone unseen. Jim, the head of the family, told us he often felt "eyes looking down on me from the second floor hallway." Betty and Tammy, twelve and sixteen, heard a card game going on in the kitchen late one night. "We heard chairs squeaking and cards being shuffled," said Betty. "But there wasn't no one there," snapped her sister. Jim added, "There's an awful lot of footsteps around here. Mostly at night. They make a *heck* of a racket walking up to the loft. And then we hear them up there. Lasts for about five minutes, or so. The cold in the living room...we can't keep it warm at night. There's no reason for it. Would you believe in one month we went through nine-hundred dollars in fuel trying to keep it warm?" Mark elaborated his experience: "I was hanging my 22 up, upstairs, when I heard footsteps in the hall coming at me. They were getting closer and closer, and then it put its hand on my shoulder. I came down the steps as fast as I could."

Everything seemed disorganized and chaotic: an environment in which spontaneous psychokinesis thrives, especially Random RSPK. Several candidates for Agent were on hand--any one of the four teenagers--but no apparent single Focus of the infestation--like effects. It occurred to me that the Bentons were, *en masse,* filling the role of "Collective Focus." RSPK sometimes reflects a psycho-environmental syndrome in some respects comparable to a psychosomatic disease--but rather than involving a single individual as does an ulcer, for instance, this malady attacks whole families.

It dawned on me that the supposed poltergeist phenomenon was completely overshadowed by the tremendous tensions in the house. I was aware that psychological and sociological tensions in families had been linked to such occurrences, but I had no idea of how dense and heavy it could be. Later when I got back in my car and drove away from that house, I literally breathed a sigh of relief.

Yet there was more to this family than met the eye. At first, they struck me as high-strung but otherwise average folks who just happened to have a mild haunting to contend with. Then they dropped their little idiosyncrasy bomb on me: The Bentons had a peculiar penchant for digging up grave sites and had been hard at it in the old cemetery out back:

Sylvia: "Our lives have gone right straight down hill ever since we came here. When we first moved in I told Jim, 'Let's go dig up the graves,' and he goes, No!' And I said, 'Yeah! I wanna see what's in the graves.' And I kept having that on my mind and having that on my mind. So one day in August, when he was home on vacation, we went out there and started digging up the graves."

Andrew: "What graves?"

Sylvia: "We got a cemetery out there."

Andrew: "Why in the world would you dig up a grave? Is that something you normally do?" (Embarrassed laughter from the Bentons.)

Andrew: "Were you just curious?"

Jim: "I wanted to see if he had old guns on him, or gold and jewelry ... you know!"

Andrew: "What did you find?"

Jim: "We didn't get that far. We got down about three feet deep and quit."

Sylvia: "The kids said, 'What are you doing?' And I said, 'We're digging up the graves, and they said, 'No. Don't dig up them graves ... well have all kinds of trouble.' They were all screaming and hollering and having fits, so I got mad and stopped. I just took the dirt and shoved it right back in the hole. We didn't do it ... the kids wouldn't let us."

It may have been their offhand denial of wrongdoing, or the seemingly lame excuse that the *children* stopped them from digging. Whatever the reason, I doubted that I'd heard the whole story. I thought I was prepared for just about anything. But I couldn't hide the disgust I felt. I speculated that they'd seen my revulsion to post-mortem pickpocketing and had made a quick turnaround in response to it. In any event, whether the remaining three feet was excavated or not, once the digging got under way the haunting kicked-up in earnest. Unfortunately, my attitude toward the gruesome pair's hobby caused them to clam up. They may have been worried that I'd turn them in.

Neanderthal man, our cave dwelling ancestor, was the first to bury his dead. Although they weren't afraid, they had a healthy regard for the deceased as evidenced by the artifacts buried-away with them. Our ancestors may not have feared them, but it's a good bet these gifts for the road were a form of primitive appeasement given in the hope that the dead would stay put. The Bentons weren't worried about that possibility. After spending some time trying to dissuade them from future macabre activities I left. I don't know if they continue to violate graveyards, but the last I heard they'd bought a Ouija board and were looking for the meaning of life; or was it the location of old guns, gold and jewelry?

The Bentons were a bit off center, but I learned something worthwhile from them. We may not always know it on the conscious level of awareness, but the subliminal mind, like Santa, knows if we've been good or bad. Regardless of whether

Jim and Sylvia actually dug up a body, the Infestation phenomena they were forced to live with were probably psychologically projected responses to the morbidly unacceptable idea of body snatching.

The Khufu Snafu

Guarding the entrance to King Tut's antechamber Howard Carter came upon a clay tablet inscribed in hieroglyphics. When decoded it read*: Death will slay with his wings whoever disturbs the peace of the pharaoh.* Carter and three dozen of his associates scoffed, but the warning proved to be all too prophetic.

On an enormously high plateau in Giza stands The Great Pyramid of Cheops. Cheops, also known as Khufu, was Pharaoh of Egypt during the 4th dynasty (2900-2877 B.C.). The construction of his nearly 500 foot high tomb was twenty years in the making. Unlike Tut, when archaeologists entered Khufu's burial chamber there was no treasure awaiting them, only an empty sarcophagus bereft of the pharaoh's mummified body. We will never know the fate of its plunderers, but we do know that the lives of 185,000 workmen were lost building the Great Pyramid--a terrible malediction in itself.

Cairo physicist, Dr. Amr Gohed, who carried out computer analyses of radiation experiments in the Khufu pyramid, told the *New York Times* that these ancient burial sites are "a mystery that lies beyond rational explanation." "Call it what you will," said Gohed, "the occult, the curse of the pharaohs, witchcraft or magic ... there is a force at work inside the pyramids that contradicts all scientific laws." (Source: *The Curse of the Pharoahs:* Philipp Vandenberg, J.B. Lippincott Co., Philadelphia, and *Famous Curses:* Daniel Cohen.) It's interesting that as late as December 1991, the speculation continued. An Egyptian scientist announced that the cause of the "Curse" was radiation poisoning coming from the slowly decaying mummified bodies.

It was observed that when Soviet Prime Minister Nikita Khrushchev toured Egypt in May 1964, he was warned by the KGB against entering The Great Pyramid, since it was believed that a prolonged stay could adversely affect his mental balance. Six months later Khrushchev was Russian history. Thirty years later, I witnessed first hand the effects of such a stay in Khufu's tomb.

In 1994 I was invited to participate in a television talk show on the topic of ghosts and hauntings. It started off very strange in that the hostess, I don't think, cared for *any* of us. In fact she made the comment (during the preliminaries) that,"I didn't want to do this *kind* of show and have you *kind* of people on here." So we were off to a good start with that. There were three people from Chicago who were involved with a ghost research organization up there. And there was another lady who does tape recordings of ghosts. The people from Chicago didn't like me at all; they thought I was a carpetbagger and was coming up there to steal some of their thunder. During the show I thought one of the ladies, who was a medium, was going to burst into tears at any moment because she seemed so depressed. She kept giving me dirty looks and dirty comments. Every time I'd say something like, "I don't know anything about that," she'd say, "Well! I didn't think that you would!"

Although most of what was said by the other guests wasn't meant to be funny, I couldn't help laughing: The hostess kept telling me, "This isn't intended to be humorous; this is intended to be serious." It didn't work that way.

And then the climax came when they had the lady on that did tape recordings of ghosts all over the world. She played her recording of King Khufu made, she said, in The Great Pyramid. In a voice that was unmistakably hers, the tape cried out:

"'OH GREAT KHUFU...MAGNIFICENT ANCIENT PHARAOH ... KING OF EGYPT AND THE NILE, CAN YOU SPEAK TO US? OH KING KHUFU, CAN YOU SPEAK TO US? CAN YOU TELL US HOW YOU ARE?'"

A long pause was followed by a tiny little voice that said, "*Just fine.*" At that point, I broke up. The technicians, who were watching me anyway ... because I had kind of disrupted this whole thing ... they started laughing. Then the audience realized what we were laughing about and they started laughing, too." By now, the show's hostess had taken on that "we are not amused" look. She refused to speak to me after the show. And there's the housewife who thought her toaster was possessed because bread would pop out with the words "SATAN LIVES" singed on it. Why did she keep the appliance? "Because," she shrugged, "when all's said and done ... It makes really good toast!"

I haven't counted them but over the years I must have received a hundred or more reports of suspected demonic interference in the lives of those who asked for my help. When I investigated, signs of it were often present but nothing more. It's doubtful that I'll ever know with certainty what, if anything of substance, molested them. I can only surmise that either my patrons have not been used by evil forces as channels to activate psychic phenomena, or I lack the means to detect that they have.

Parapsychologists are light-years away from learning all there is to know about hauntings. Emerson said, "Every man is haunted by his own daemon." Since we lack proof that directed evil exists it's easier and more comforting to believe that my suggestible clients are conjuring up their own daemons. Yet, at the risk of appearing the fool, it is difficult to consider cases like these as self-inflicted misfortune without at least speculating on the role devilish influences may have played in them.

Chapter 10: Into the Unknown

"*Once men are caught up in an event they cease to be afraid. Only the unknown frightens men.*"
Saint-Exupery, *Wind, Sand, and Stars*

Case File: The Haunting of Barrington House

Nell: "Good God, it knows my name! It knows I'm here!"
(*The Haunting of Hill House,* Shirley Jackson)

Nature's little nuisances: bats, squirrels, a mouse or two--creatures that trespass the stillness of the attic--these we expect. On an evening in May 1992, I summoned an intruder to the garret of Barrington House that some believe was beyond nature. In *Four Past Midnight* (Stephen King, 1990), the prolific author tells us: "Writing is an act of self-hypnosis, and in that state a kind of total emotional recall often takes place and terrors which should have been long dead start to walk and talk again." When I recall that night, hammering fear mixed with the exhilaration of discovery come flooding back. Barrington was a plunge into the unknown-and it rocked me!

A decade has passed since I investigated the old mansion; but even now, long after coming to terms with what happened, the ghost that walks its halls continues to plague me. Resembling many of the troubled houses I visit, Barrington was, and probably still is, a mixture of the real, the imagined, and, no doubt, the overstated. As far as I had the opportunity to discover there wasn't much in the way of spooky stuff going on there. A few disquieting moments, to be sure, but the *haunt* was and still is fairly typical of all energy patterns. My difficulties with Barrington centered on a disturbance-ridden seance, not on the fact that it was ghost-ridden.

The following is the full and accurate saga of the old mansion; or, more precisely, as much of it as I was allowed to uncover during my brief stay. I have reconstructed the events using tape recordings, newspaper articles, extemporaneous notes, interviews, and my recollections. I have leaned heavily on quoted statements made by the participants.

Barrington House is an old brick and stone mansion in Philadelphia, Pennsylvania. It rests side by side with other ancient edifices put up just after the turn of the century. Apart from the "ghost," its only claim to fame is that it's situated on the site of a Civil War era military hospital. History's clock started running on the imposing place in 1912, when Lawrence Edmond had it constructed for his daughter Ida and her husband Jeremiah Parry. At the time, Edmond's other daughter, Emma, and her husband, Clinton Barrington, lived next door in another of Philadelphia's architectural wonders (currently serving as church rectory). It wasn't until 1928, when a member of that clan moved in, that the structure could be properly called

Barrington House.

Apparently, old Lawrence Edmond so admired his gift that after construction he and his wife, in the tradition of old-world Europe, moved in with the Parry's. The Edmonds resided in quarters at the east end of the second floor. Since there were no little Parry's running around the sprawling mansion, it must have been a quiet existence for the four of them.

When Ida Parry passed away in 1924, she willed the house to her niece and two nephews--her sister Emma Barrington's children--with the proviso that husband Jeremiah could remain. Jeremiah

Parry passed on four years later under somewhat peculiar circumstances, which may or may not have had a bearing on the subsequent events, Clinton and Emma Barrington's son, John, purchased the estate from his brother and sister in 1928. John lived there until his death in 1960. Then, in 1976, having no progeny of her own, John's widow donated it to the Catholic Church and moved to the Imperial Hotel at the opposite end of Philadelphia.

Remodeling began in earnest a year after that. Almost immediately there was chaos. By December 1969, when the newly converted structure opened as an office building for Catholic Charities, Barrington House was already showing signs of being a very unusual place... It was Marion Zarkauer, preeminent Philadelphia historian, instructor at the University of Pennsylvania, lecturer, and author of books and articles on the area's heritage homes that brought the haunting to my attention. Mrs. Zarkauer, one of the most energetic women I've ever met was, and still is, keen on the subject of "things that go bump."

It was, I believe, sometime in 1988 that she originally thought of writing an article on Barrington and on the woman who had so generously given the Church a home. As her interest in it grew, Marion and Edith O'Brady (a volunteer worker) took the opportunity to interview the aging widow. Later, in a December 6, 1991 telephone conversation, Marion told me that Mrs. Barrington (now deceased) had been "terribly interested in the idea that I might do a little write up about it." She continued: "Talking to Tom one day about it, and he thought it would be kind of interesting." (Tom Fletcher, a member of the managing board of Barrington) But she later decided that if she could track down a spook it would make this article more interesting.

Although direct witnesses to the haunting were not easily located, she eventually turned up some interesting tidbits. Fortunately, she got them on tape. The witness, a housekeeper for the Catholic Charity, indicated that it was mainly footsteps and voices that were heard:

Marion: "Just this last week, or maybe the week before ... it was something in the attic and two people heard it. Then they went up to explore, but there was nobody there...Primarily, hearing footsteps and voices. And now the angry voice is excited. Occasionally, apparently, the anger becomes very dominant. They tied a lot of it up with the pool-the goldfish pool out here. Now it was covered up and they felt that this caused a lot of commotion: the angriness and feeling of turmoil, particularly during that time."

Andrew: "You say *angriness.* You mean while they were closing up this pond?"

Marion: "Yes, there was. As I say, they found something ... *terrible.* And they related that, somehow or other, to the feeling that somebody was terribly disturbed about this. But somebody is looking for something! One of the first things they heard were footsteps."

Andrew: "Now, where were these located?"

Marion: "These would come up the stairs, and they would always be in the old section; always in the east end of the hall, where the man who built the house (Edmond) had his suites." At this point, Mrs. Zarkauer recited a brief history of the house; the identity of the builder, and the fact that the other Edmond sister, Mrs. Barrington, had. lived next door for many years. She paused for a moment, then resumed:

Marion: "First it was footsteps going down the hall. Then it was hearing voices: these angry voices,

particularly at the time when the pool out in back was being taken care of. Then it actually was somebody trying to trip this employee Kathy. She was coming down the hall and somebody actually tried to trip her ... and she almost fell."

Andrew: "Now, how long has Kathy been associated..."

Marion: "She's been here since the beginning...Pretty close. Shirley (another employee) and Art (Arthur Kelly, the manager) were the two people who back, let's see maybe in July of '69, when the church was still renovating the building ...I mean it wasn't opened yet, but we'd hired *them.* And Kathy came, I think she was here before we officially opened in December 1969. She was here ... probably October."

Andrew: "Well, did she begin to notice something immediately, or was there a period..."

Marion: "I think it was from the very inception, but it took a long time before they were willing to admit that this was what it was. But they talked to one another about it very much."

Andrew: "What is he experiencing?"

Marion: "He's experienced *everything.* He's really more...almost more interested than anyone because he's been touched."

Andrew: *"Touched?"*

Andrew: "Yes, *he* was, too. Now he was not tripped, but he was touched, too. And Kathy, apparently sitting at her desk. .. and somebody actually pulled on her leg, she said, actually grabbed her ankle. "And then I think one of the oddest things is when she was in the bathroom that time and the door opened and somebody actually walked in."

Andrew: "You mean she could sense somebody there without actually seeing them?"

Marion: "Uh-huh. And the door actually opened up and footsteps actually came in. And usually they are men's footsteps Now whether they were coming into the women's toilet ... you know, that would upset her. I forgot to ask her that part."

Andrew: "Well, one thing ... we'll check with her to see if they fall into any particular pattern. It will, or should be doing pretty much the same thing all the time." (I was already anticipating an Imprint in the ointment.)

Andrew: "Because, if he's looking for something, it'll follow the same pattern almost identically. More than one person observing the thing should observe the same pattern. Now I say *he* because..."

Marion: "Now, *she* (Kathy) says they're men's footsteps, and the voices are men's." "And the quarrels are between *men."*

Andrew: "They hear more than one voice?"

Marion: "There's *quarreling!"*

Marion: "Andrew, if you were upstairs now--you know, after lunch--and would sit quietly in a room ... would you be able to get an impression?"

Andrew: "It's quite possible. But see, I hate to get anything started that I'm not going to be able to finish. You've got a time element here. If I do an investigation, first of all I'd like to get other people here as witnesses. I'd like to be in a position to carry it out so we can see where we are." "Now a very valid question at this point is, Do these occurrences really disturb the people?"

Marion: "No, not any more; they've gotten so used to it. In the beginning they frightened them ... upset them, but they've just gotten used to them like a lot of people (who) go through these things. Sometimes they do frightening things, too! Some of these old houses ...I don't see how they stand it; it's *terrible* what some of these ghosts will do. They just scare the beejeepers out of everybody. I don't know why the family stays there." (A question I have often pondered.)

Exactly five months later, on April 19, 1992, I returned to Barrington with two assistants: Philadelphia Psychical Research Society members, Terry Hatley and Bruce Lund. Also in attendance were Marion Zarkauer, Arthur Kelley, and Kathy Simmons. Again the proceedings were tape recorded. I opened the discussion at 9:15 p.m. By virtue of the fact that she had been closest to the haunting, Kathy Simmons had center stage:

Kathy: "When we first came into the building, Mr. Kelley and myself was the only ones with the construction men; and for about two months we both had feelings that there was something in the house, but we didn't tell one another ... you know. We were kind of, not ashamed, but we just thought it was our imagination. I mean, I'd be coming in here and I'd definitely feel something hovering, and he got it, too.

"And then we were finally opened, and we were on our own. And I'd be up here in the morning by myself, and I'm not a scary-type person. I believe in the spirit world, the supernatural. I'm a believer, and I'm not afraid of anything. But I kept hearing ... I'd hear these heavy footsteps all the way down the hallway, and it would go down the steps, and sometimes he would come up the steps, and he was, like playing games with me. I'd come out in the hallway and say, 'Yes, may I help you?' and, you know, there'd be nothing there. And this was the games he would kind of play. It was like a game.

"And I made my desk a certain way ... locked ... and the cabinets locked, and I'd come in the next day and it would be messed up, and I'd say, 'Mr. Kelley, were you in my office?' I don't have anything to hide, but I just leave things certain ways and there's certain things Mr. Kelley doesn't want the other employees to see. I said, 'Well, does anyone else have a key?' No, no, no! Well, as soon as we both admitted that there was something here, these little games stopped.

"So spring came, and we both ... oh we'd hear the walking and the slamming of the doors down the hall, you know. *SLAM!* you know, would go down the hallway. This was like to get your attention. And the doors would be opened and there'd be nothing. And we weren't afraid of it. I was at first, 'cause I felt that it was menacing in a way sometimes. So every time I would feel its presence, I'd think, you know, *I love you and you're a thing of God and I love you... and He does, and I mean you no harm. 1 love your house. I'd love to know you, and can 1 help you?* And I would think, like, good thoughts. And it got better.

"So then they decided that out here in the back there was a fish pond and it was buried, covered up, but the concrete was still there. So they decided that nothing would ever grow out there unless they dug up this fish pond; drilled it out and then filled it in again. So, they started doing that. Well he ...I think of it as a man ... got very violent at this period of time. You would go downstairs and the doors would shake. And one time I went down the steps and it grabbed my ankle right here (indicating the spot). I was going down the steps and it nearly made me fall. And the impression of him grabbing me was there for a long time, just like a strong male had just grabbed hold of my leg.

"Well, we'd hear arguments, just *horrible* arguing in the lobby. Mr. Kelley and myself both would go running out there and say, 'Who is it? Who is it?' Nothing would be there. You'd go out into the lobby ... nothing would be there but this horrible, violent quarrel.

"When the fishing pond was completely dug up and returned to the way it was originally, it calmed down and it's presence wasn't felt for, you know, a couple of months. But then it came back, and it has been very passive since then. But, I mean ... it's here!

"So we got this new old German chef and, of course, almost every employee that works here believes in the ghost: they have felt it; they have heard it; they have sensed it in some way. Well, Mr. Kepler didn't believe. So one afternoon he was up here and *it* was walking. And I had gotten to the point that I just didn't even look anymore because, *it* was there; I knew *it* was there; he knew I was there, and you get tired of (the) games. So Mr. Kepler kept saying, 'Why doesn't Kathy come into the office? That is Kathy (walking)?' Mr. Kelly said, No that's not Kathy, that's the ghost'. No, no, no!

"So anyway, they finally got convinced and they both walked down the hallway into the bedroom; walked around in the bedroom. Mr. Kepler did hear the footsteps (he) slammed the door, and he goes, WE GOT IT! WE GOT IT!' And he goes in there, and there's nothing there. And, then, Oh! He curses and he's slamming. And he said, 'I've heard it, I've heard it. .... and it's making a fool of me.' Well, *of course* it was making a fool of him. It likes to do these things." Kathy continued: "We had a girl...Jane Dover, my very dear friend, very close friend, and she didn't really believe too much. And one night she came up to change clothes to go home. It was about this time of the evening; and she thought, *Oh, I'll just use the rest room since nobody's up here.* So she went into the new section and went into the bathroom ... and she didn't think she was going to get out. It was playing games with her, and it was holding the door, too. She was pulling this way (yanking hard on an imaginary handle); it was keeping the door closed. When she finally got out of there she said she ran, and she never came back.

"Now one day last week, I was sitting here (in her office). It was the morning again, and I looked out and there was the definite impression of a man just turning that corner (second floor hallway), enough so that I got up and I thought, *Well now, who is this?* And I went out into the hallway and I said, 'Willie? Chris?' You know, 'cause that's our two boys that are here in the mornings. There was no one."

Andrew: "You said you had the impression ... of what?"

Kathy: "Well, I thought I might see something, you know. But I know it wasn't there."

Andrew: "You couldn't distinguish, truly, what it was?"

Kathy: "It was a male, but I mean I couldn't describe him at all. I didn't know who it was, but something definitely was coming around that corner."

Andrew: "Well now, is there a particular pattern, Kathy, to these things? Any particular time of the day?"

Kathy: "No. It's very active in the spring. Uh-huh."

Andrew: "In other words, it can happen during the day."

Marion: "Oh, it happens anytime. Yeah. Day, afternoon, it knows no limits. Most of our employees refuse to go to the third floor (the attic)."

Andrew: "Why do they do that?"

Kathy: "They're just afraid. I guess when they get strong feelings upstairs they don't like to go up there."

Andrew: "Where is that?"

Kathy: "Over-right up those stairs; over there by the ... the entrance is down there by the ladies room."

Marion: "What was it used for?"

Kathy: "Servants lived upstairs."

At this point in the discussion, Ms. Simmons was asked about the gender of the ghost again, and she replied, "I really couldn't say definitely, but I've always felt all along it was a man." The question of violent quarreling without an observable source came up. She hesitated a moment, then answered in a steady voice, "Mr. Kelley and I really heard the arguments. It was enough so that one Saturday morning Mr. Kelley came out of the kitchen with a knife, because he thought that there was people going 'to if in the lobby and he wanted to break them up." I asked if either of them recognized the voices. She shook her head: "No ... Mr. Kelley and I have never been able to. And I've asked it ... you know, Tell me!' you know, 'Help me!' I've tried to communicate with it but it doesn't. Maybe I don't know how to."

Overseeing the $350,000 renovation of the Barrington and its daily operation was Arthur Kelley, manager of the house from 1969 until his retirement in 1991. He has been described as a level headed businessman. In the October 26, 1992, feature article, "The Ghost of Barrington House," Kleine told author Zarkauer that when a landscaping crew dug up the dirt filled ornamental pond behind the mansion they uncovered a doll and a baby carriage in the process. Previously, he reported that when the excavation got under way, "Boy, was there some racket in this house. (Initially)...there were just little things ... the kind of things that were hard to pin down. But they happened so often that I began to get this eerie feeling, like I wasn't alone."

What kind of things? Tools that had been set down by workmen moved by themselves, or simply disappeared. Doors kept slamming shut, especially the door to the third floor attic. No matter how often they propped it open it would invariably be found shut tight. Unexplained footsteps were heard and other odd noises, causing feelings of uneasiness. One crew member was said to have been so unnerved by the goings-on that he quit. Kathy Simmons told me that the workers felt closing the pond had caused much of the commotion. As she put it there was "angriness and feelings of turmoil..." Some believed these hostile emotions were emanating from an unseen presence within the house. Sounds without a source would frequently echo throughout Barrington; at least once they escaped to the street. It was common knowledge that people walking on the street nearby heard the uproarious sounds of a party coming from the mansion; except there was no party going on in the completely darkened building that night.

Arthur Kelley experienced more than his share of unexplained sights and sounds. "After unearthing that baby buggy, all kinds of things started happening in that house. Sometimes when I was alone, I'd hear footsteps go up the stairs to the second floor, but when I'd go up to see who was there, I couldn't find anyone." Later, Kelley told a Philadelphia newspaper reporter, "I would be working in here alone about 10 o'clock on a Saturday night, or even a Sunday morning, and I swear I'd hear voices, or footsteps, or something, and all the doors were locked. I used to take this (a thick glass rod) and go walking around the house, but I would never find anything. One night I was so sure someone was in this building, I called the police. Two armed policemen and myself walked all through the building and didn't find a thing..." Kelley said this happened about two dozen different times and usually when he was working quietly in his second-floor office: "I would go home and tell my family, and my wife would say, 'You're hearing things,' and the kids would say, 'Aw Dad, you're gettin' old.'"

Kathy Simmons was not the only observer of things ethereal. Kelley was "made" to see it too. He told me, "I was standing right here (just outside his office) when I saw it. That light right there (pointing to a chandelier about 15 feet away) was on and down the hall came this mist towards me. It was like a big puff of smoke coming toward me--about five feet tall. And then it got about 10 feet away and dissipated." "It's a bad analogy, but it looked like room deodorant spray. I rubbed my eyes, thinking they were playing tricks on me. It was still there." Kelley described it as "...a formless, yet somehow discrete vapor that moved slightly and disappeared." Unlike Simmons, Kelley was unable to assign his vision a gender. Kelley, a non-smoker, said the building had been closed for about two hours. He's not sure what it was, but he did see it. It wasn't enough to make him a full-fledged believer, though, "I'm still not sure if I believe in this stuff or not." "There were a few times when I thought I saw a thin mist or fog moving down the hall on the second floor." I wanted to know, "Why did he never say anything about it?"

Kelley: "People laugh. They say, 'Look at this guy, what's he talking about?'"

Simmons: "But Kelley definitely sensed a presence in the house during the fifteen years he worked there."

Kelley: "Something was there; I didn't know what it was. It wasn't harmful, and it wasn't bad. I've never experienced anything like it (in) any other building in the 58 years of my life."

The most dramatic phenomenon Kelley experienced was the phantom argument in the lobby. After a large, fine party running well into the night, Kelley stayed late to close up.

Kelley: "I heard the noise of a big fight going on. There were loud, angry voices in the lobby, like several men yelling. I went into the hall and everything stopped. No one was there. *No one!* Then I

heard it again. I thought I had left the front door open and some partyers had come back into the building. I walked back into the kitchen, went to the chefs table and grabbed a big knife out of the drawer. I was carrying the knife with me. That is how sure I was that there was something there. It scares me now. I get goose bumps thinking about it, and that was nearly twenty years ago."

I Take the Case

Genuine hauntings are not the result of an occasional odd sound, a pungent aroma, inexplicable flash of light, or an eerie feeling. Haunting experiences are composed of a complex series of events; a systematic chain of uncanny happenings that lead the occupants of the place to wonder what on earth's going on. Barrington is such a place! Public buildings that have been "touched" are an enigma. Not only are they less interesting to the media as story lines, they're more difficult to research than private homes. You see, no one lives there on a regular basis so they lack most of the human element: the warmth and the emotional and intellectual climate that seems to permeate private dwellings. Phenomena witnessed in these sterile surroundings usually lack continuity. There's seldom a single Focus for me to latch onto. What's more, attempts by an alleged spirit to communicate with those who temporarily occupy the building rarely take place. Even if they wanted to make contact, most public structures, including Barrington House have been emptied of their human "receptors" by the time the bewitching hours arrive.

On the other hand, although these short-comings may discourage spirits and demons, they're pretty much characteristic of the haunting grounds of Imprints. As we have seen, many so-called "ghost-plagued" buildings play host to these energy patterns. Why go through the motions of a Psi Session once the case had been diagnosed as an Imprint: a category beyond the "healing" range of this simple form of exorcism? The answer lies partly in the fact that in those days I was hell-bent to take any case I could get and eager to use the entire investigative process--including the séance. And yet, it was much more than that. Arthur Kelley's remarks, coupled with the direct testimony of Kathy Simmons, were most compelling. Their statements were reason enough for launching the hunt.

And let's not overlook Marion Zarkauer. Apart from the fact that she was a fellow Society for Psychical Research member, her position in the community as its official historian regularly brought her into contact with people who had need of "ghost hunters." Taking the case as a favor to her might have served as an inducement for future referrals, if only I could have ended it on a more successful note. That, unfortunately, was never to be. Years later, further corroboration of the haunting of Barrington House surfaced. Statements given the press subsequent to my traumatic visit confirm that remarkable demonstrations continue in the old mansion.

In 1987, Bill Trask, an employee since the transformation from private home, said that after hearing footsteps late at night coming from the third floor attic he believes the house is still haunted. Margie Bell, a secretary, agreed with Trask as far as the location of the disturbance is concerned. She told me that she has heard voices, both male and female, arguing in the night. According to Brenda Snelling, another secretary for the Charity, "I did experience something two years ago which makes me think that maybe there is something mysterious in the house. It was Christmas season, 1985, and Snelling was alone in an office upstairs":

Brenda: "The steps were creaking, like someone was walking on them. I went out into the hall and looked, but there was nothing there. I heard them again and finally I got up and locked the door and said 'to hell with whatever is out there.'"

In 1989, Shirley Emerson was managing Barrington. Shirley's not a believer but offered this story to me: "One of my volunteers swore that it was down in her kitchen in broad daylight. It was the top half of a man standing there looking at her. She is not the kind of person given to that nonsense." I also interviewed Richard Evans, associate director for business affairs at the Catholic Charity, and another skeptic:

Andrew: "You also confirmed that the house is said to be haunted? "

Richard: "There are many employees there who still swear that there is something there. A lot of them are scared to go up to the third floor."

And there's Alberta Jackson, housekeeper:

Alberta: "Sometimes when I'm cleaning downstairs, I feel--like a wind--passing by my arm. You wouldn't catch me up there on the third floor."

Andrew: "You're not frightened enough to quit, though."

Alberta: "I figure I never done nothing to it, so why should it do something to me?"

In 1989, Arthur Kelley told me that before my investigation he "took his account of events in Barrington House to his supervisor in the Catholic Church who stiffly suggested that he 'keep quiet and not make waves.'" Understandably, after my visit the Church brass continued to shun sensationalism: Kelley: "When rumors of the haunting began to circulate in the community in the early 1970s, the Church was less than eager to discuss them publicly. Although detailed reports appeared in the local newspapers, reports were curiously sparse in the Catholic press. Some have attributed this to the efforts of the Church's public relations man at the time, who pleaded that such publicity would damage the Church's prestige."

To this Ghost Detective, heading the list of bewildering circumstances surrounding Barrington was the atypical action of the Church's administration in sanctioning my visit. The riddle was solved, however, when I started researching material for this book. The fact is I never had official approval to investigate! Approval came, not from the chancellor's office, but from managing board member, Arthur Kelley. As far as I can tell, and I don't mean to supply his motive, but it was apparent that Mr. Kelley had been prodded into action by that frequent observer of phenomena–Kathy Simmons, and by the "ghost-conscious" Marion Zarkauer.

During the December 1991 telephone interview–referring to her planned article–Marion eagerly acknowledged, "if we could track down a spook it would make this more interesting."

Apparently, Mr. Kelley hadn't been too concerned about the consequences--the possible downside to putting me on the track of that "spook." Arthur Kelley who, together with Ms. Simmons, appeared to be the Focus of the haunting would have been one of the prime time players at the investigation. He told Marion Zarkauer that he asked Kathy to attend the seance in his stead, "because he had other plans for the evening." But he didn't. He simply chickened out. In the Fall of 1992, I got round to asking Marion Zarkauer why she didn't attend that night. From what she told me, Charles, Marion's husband, had advised against it. Apparently, he thought it would be better if
she stayed out of it, in as much as it might prove detrimental to her career.

So, much to my chagrin, opening night at the Barrington consisted of only four intrepid souls. At eight forty-five, on the night of April 19, 1992, Kathy Simmons, psychic investigators Terry Hatley, Bruce Lund and Andrew Nichols climbed the narrow steps leading to the third floor attic; snapped on the bare light bulb just outside, and did their best to prop open the door with the propensity to slam shut. Why the attic? Kathy wanted to have it in the lobby downstairs. She thought it was rather strange that I insisted on doing it up there. But the third floor attic was the one place most employees seemed terrified of (and still are for that matter). Arthur

Kelley said that heavy footsteps often thudded in rooms above his second-floor office. And there was the ladies' room located near the stairwell leading up to it. Both Kathy and her friend, Jane Dover, reported unnerving experiences while in the rest room; and, of course, there was the obstinate attic door. Another incentive for going upstairs was the fact that it was totally isolated from the rest of the building, virtually eliminating the chance of our being interrupted in mid-experiment.

Why not the lobby downstairs? Except for typical psychic patterns: the sound of partying; of several men "going to it" in an argument; and yes, footsteps and doors slamming, too--there really wasn't enough going on to warrant a séance there. What about the second floor hall or Mr. Kelley's office? Footsteps had been heard padding along the carpeted hallway and misty visions seen, as well. There were a number of promising locations, but we could set up in only one (although, looking back on that evening, I believe we would have moved to an alternate location had the attic been less productive).

Outside, the night was still, warm. Inside, the atmosphere was stuffy and thick with the dust we'd churned up. Soon it was dark enough for candles. They flickered, casting ominous shadows of us on the attic walls. It was probably my imagination, but they seemed to use up more than their share of the stale air. We forced open a window, then took our seats at the card table from Sears: Kathy across from me, Bruce opposite Terry. I remember my surprise at their composure. If either young lady was apprehensive it didn't show. While my adrenaline was surging, they had only expressions of expectancy.

Kathy was first to operate the pendulum. She quickly established the movements for the responses Yes, No, and Unknown. The following are excerpts from the tape recording we made that night:

Andrew: "I ask you now to get an answer from the pendulum to the question ... Is there an unseen presence in this house? Let's see what kind of response you get. Ask it out loud."

Kathy: "Is there an unseen presence at work in this house?"

Andrew: (Simultaneous with the movement of the pendulum.) "You're getting Yes. Is there more than one presence?"

Kathy: "Is there more than one presence in this house?"

Andrew: "No, on that one."

Kathy: "Uh-huh. Did it mean us harm?"

Andrew: "You got a Yes response on that."

Kathy: "Uh-huh. Was it because of the fish pond removal?"

Andrew: "No! That's kind of interesting."

Kathy: "Was it because of the changes in the house that we made?"

Andrew: "Hmm...no answer."
Kathy: "Did it just resent us being here?"

Andrew: "No. It can't be that."

Kathy: "Are you the presence that I feel so strongly in the morning? Are you a male?"

Andrew: "Yes!"

Kathy: "Uh-huh. Do you realize that Mr. Kelley and myself and all that work here feel kindly toward you? Did you die in this house? Uh-huh."

Andrew: "Yeah."

Kathy: "Was your death sudden? Were you ill for a very long time? It's kind of a weak No.... I don't know. Did you die suddenly?" (No response.)

Kathy: "Were you responsible for your own death? Did you kill yourself?" (A definite Yes came through.)

Terry Hatley: "That's a very strong one."

Kathy: "Uh-huh. Well, were you a Barrington?"

Andrew: "Nothing!"

Kathy: "Well, let me ask it again."

Andrew: "That would explain a great deal."

Kathy: "Were you a Barrington? (Long pause without a reply.) Are you a Barrington?" (No response.)

Bruce Lund: "It might be by marriage. It could have been a girl."

Kathy: "Were you born a Barrington?" (No response.)

Kathy: "Are you dead?" (Strong but indecipherable response.)

Bruce Lund: "I don't think he likes that too well."

Andrew: "It doesn't recognize it's dead, I guess."

Kathy: "Uh...Are you unhappy with your present condition?" (No response.)

Kathy: "Are you disturbed with your present condition?" (No response.)

Kathy: "Would you like your condition changed?" (Nothing.)
Notwithstanding the fact that the responses were of no evidential value, I remember thinking that Kathy possessed either an extraordinary intuitive ability, or the "smarts" to know exactly what to ask. With the exception of the first few, neither I didn't supply the excellent questions she was asking.

Kathy: "Should I ask him if he'll talk to us?"

Andrew: "Yes."

Kathy: "Would you communicate with us? " (No response.)

Kathy: (With a trace of amusement in her voice.) "Sir, would you *materialize* so that we can see you?" Her request for a personal appearance triggered the first of the evenings' uncanny events:

Bruce: "Now don't say anything, but I think that light out there's gone off (the bare bulb hanging beyond the open attic door).

Terry: (In a small wavery voice) Has the light gone out in the hall?"

Kathy: "Let's check it out." (A moment later) "The lights on in the hall."

Bruce: "'Thank God!"

Without hesitation, Kathy left us to find the bulb still burning. The door that had been carefully propped open, however, was now closed tight, cutting off all the light, except from around its edges. Barrington's persistent doorman was on the job.

We were getting nothing from the pendulum, so turned to the card table. It was my intention to continue the questioning by way of table rapping. But the little stand was psychically primed and quickly out of control. No sooner had we placed our hands on its top than it lurched, rocked back and forth and took off' It was all we could do to keep up. At first, two of us–Kathy and I--ran at its side; Bruce chased from behind, while Terry was pushed backwards across the attic floor. Whenever we lost touch with the "runaway" it would stop dead in its tracks, wait for us to catch up, then take off again at gallop speed.

Once it came to an abrupt halt, raised *off the floor* on one leg and did a pirouette--then slammed down hard and began running again. A little later, the tubular steel leg on my right lifted off the floor, the remaining legs staying put. It got up so high that the corner of the table became a creased triangle Then it came crashing down with a force that bent the leg nearly in half. I'd be walking with a cane today had it landed on my foot. The passage of time exacts a toll on our memories, yet I never lose the excitement of that moment. The air was alive, fueled by imagination; by uncontrollable fear; and something else. We saw nothing, but I was convinced that something had joined us and was watching. I was sure my companions felt it, too; but we just looked at each other in dazed disbelief.

The tension was unbearable. One of the girls (Terry, I think) started giggling. For my part, I did the best I could to hide my feelings. I tried to appear nonchalant--as though these things happen all the time. Underneath, I was shaken. The panic of childhood came flooding back. I was a small boy again, reliving nocturnal terror; fantasizing shapeless forms in dark corners of my bedroom. But I was no child, and this was no imagined bogeyman. I was wide-awake observing a dimension of energy that was gradually taking control... We were no longer conducting the ceremony-we were following its lead.

I'd been prepared to spend another "ho-hum" evening waiting for something to happen. Nearly everything I knew about spectral phenomena guaranteed our safety on the job. At the moment, I wasn't all that convinced. After our card table did its version of *Dancing In The Dark,* our foursome resembled a deathwatch as we sat waiting for the inevitable. I was losing my objectivity. A voice whispered in my mind, *Are you sure you should be here?* I glanced across the table at Bruce. He looked at me as his Van Helsing: archenemy of the supernatural. *If I lost my cool,* I thought, *how could I hope for them to keep theirs?* But his expression gave him away. He looked shocked and winded by the chase; we all were. And so, apparently, was the card table as it sat motionless on the wide slatted floor. We took advantage of the lull, approached the four-legged beast the way a bronco buster sneaks up on an unbridled horse--then slid into our seats to work the writing planchette. Although perhaps nervous and weary, neither Kathy nor Terry objected. Some bizarre things had happened, but, to the best of my memory none of us wanted to pack it in--not yet, anyway.

During the opening five or ten minutes a good deal of scrawling gibberish came through; then the planchette took off much the same as the table had. As long as hand contact was maintained it flew across the pad of paper, stopping only when our fingertips slipped off, or when the tripod hopped onto the table. It paused momentarily, then with powerful strokes words began to appear. The pencil dug deeply into the thick newsprint. At first, the unsteady pool of candlelight spotlighting the pad made it difficult to make out the letters. Then, suddenly, they became all too clear. The planchette had slashed-out the warning "GET OUT OF MY HOUSE!" followed by the horrifying message "DEATH TO KATHY!"

All unholy hell broke loose! The windows that had not been opened, flew open with a bang! An icy wind swept through blowing out the candles and plunging us into blackness. We all screamed. I could barely hear my half-hearted assurances over the screeches of the others. Kathy was in a blind panic. I fumbled for the matches for what seemed like an eternity, then re-lit the candles at last. The girls pleaded they'd had enough; so had I. We couldn't have removed ourselves from the premises any quicker if we'd have *jumped* out the windows.

Later that night, I took a look at the card table. The hollow metal legs were still in good shape, which was a surprise considering one of them had doubled in half. I'd either imagined the damage or the table had miraculously healed itself. It took another week for the repercussions to catch up with me. Marion Zarkauer told me that the two young ladies had been seized with terror during the ordeal and continued to be badly disturbed. Later, I heard that Terry refused to sleep in her bed; and Kathy, too scared to return to the office threatened to quit her job. (Eventually, she did leave, transferring to the Washington D.C. Catholic Charities office.)

Arthur Kelley was furious with me. But it was James Hatley, Terry's stepfather, who was the most steamed. Marion said he was livid; that he would have filed charges against me if he could have somehow avoided the publicity that such an act was bound to generate. My crime? The consensus was that Bruce and I had tricked the girls. We had rigged the attic and created the illusion that Barrington's ghost was after them. Nothing could have been further from the truth! I was shocked by the accusations, but understood the reason for them. To the uninformed observer physical demonstrations during automatisms are hard to swallow if not downright unbelievable. It was far more reasonable to believe I'd hoodwinked them than allow that what happened that night might have been genuine: not necessarily supernatural, but genuine all the same.

It's often easier (as well as more comfortable) to account for the inexplicable by explaining it away. I've always considered a knowledge of stage magic indispensable to parapsychologists. Not only is it a precaution against fraud, it is an equally valuable tool when looking for natural explanations for otherwise baffling events. I sometimes use this knowledge to spot a well devised hoax: as a defensive, not offensive tool of research. Houdini himself would have been hard pressed to create some of the remarkable effects we observed in Barrington's attic. I would have needed some pretty hefty technical assistance just to attempt them. I had no such help. I am not in the business of bamboozling clients. Moreover, if I were in it for the publicity, why didn't I bring someone from the press with me? And let's examine my reason for going to Barrington in the first place. I didn't just show up on their doorstep; I was invited. I was called in--after the Church had kept a lid on the story for so long--due primarily to the urgings of Arthur Kelley and Kathy Simmons. I was let in on the "secret" at their behest. If they hadn't talked it up, I seriously doubt that the Church would have consented to the meddling of a psychic researcher.

There are three qualifications to look for in a psychic investigator: reputation, reputation, reputation. You can imagine the damage Ms. Simmons and Ms. Hatley did to mine. To be as magnanimous about this thing as I can it appears two among us is amnesic. How else explain our divergent points of view. Kathy claims there were no apparitions, and no one saw anything. In light of the other witnesses, and her own tape recorded statements, how can this be? Her testimony was witnessed by five people, including representatives of the Catholic Church, so her denial of the events is somewhat of a mystery.

It may be reaching a bit, but one possible explanation for the gaps in her memory is "retrocognitive dissonance" *(retrocognitive* meaning prior knowledge, and *dissonance,* inconsistency). Earlier, I mentioned the tendency of haunting victims to forget unpleasantness, or selective events that are not consistent with logic. Used in this context, it is the slightly modified theory of psychologist, L. Festinger (*A Theory of Cognitive Dissonance,* Row, Peterson, 1957.) It occurs when witnesses to hauntings finally become aware of the wide discrepancy between the uncommon event and common sense. The difference creates anguish, and they become unconsciously motivated to do whatever is necessary to reduce the resultant conflict. The most likely method is to rationalize the inconsistency away: to put selective elements of the conflict out of their mind.

Retrocognitive dissonance is well-known to parapsychologists and accounts for the doubt that grows within many; a doubt that focuses on the credibility of their own senses. Another explanation, one plain and simple, is the fear of losing her job. I'm sure the Church didn't take kindly to her participation in the seance. It doesn't make being called a charlatan any more palatable, but it does explain Kathy's position: between solid mineral matter and a rigid portion of space.

As for Terry Hatley, Marion Zarkauer tells me she's living in Pittsburgh. Terry remains adamant about the seance, agreeing with Simmons that we tricked the two of them. However hard it may be for them to believe, we did not produce the phenomena that night, no more than we were responsible for all the other haunting effects that occurred both before and after my one and only visit to the Barrington attic. Looking back, my involvement with the case-composed as it was of equal parts of mystery and controversy--was a mistake. I should never have performed the seance without at least one official representative of the Church on hand; if, for no other reason, than to referee the dispute. Yet, in spite of the shortened ceremony and the unpleasantness that followed, I've succeeded in coming up with a reasonable solution to the haunting.

After a decade to think about it, I feel safe in saying that Barrington played host to two completely different psychic events. The first, and most persistent, was a mindless energy pattern: an Imprint featuring the sounds of muffled footsteps; of slamming doors and windows; of late night parties; loud arguments; and even the appearance of misty and misshapen forms. This ghost still stalks the halls of Barrington! The second took the guise of a singularly menacing presence. Unlike the Imprint, this ghost required flickering candlelight and subliminal devices to manifest. When it did, we sat (and sometimes ran) spellbound while a table with a mind of its own ran rampant; while dormer windows banged open by themselves; and "death," in the form of a morbid threat was stabbed-out on a pad of paper. It was a scene right out of the movies.

"The supernatural is the natural not yet understood," said Elbert Hubbard in

The Note Book. If what I suspect about the underlying cause of hauntings is true, what happened in the Barrington attic that night was not outside the limits of Nature, just beyond our understanding of those limits. The brief havoc we witnessed was almost certainly created and projected by a living person. That person no longer walks the halls of Barrington! I do not mean to imply that Ms. Simmons is a physical medium. My point is, simply, that if my adventures have taught me anything it is that where physical phenomena abound, a living catalyst is on hand to account for them. There is a theory that everyone is born with the potential of extended sensory perception; some more than others. Unless this natural gift is developed and practiced early it diminishes with age. Left undeveloped, basic mediumistic abilities such as spirit communication and the projection of mind over matter, may wane somewhat, *but they remain potent.*

I like to reassure my clients--those who wonder if they've gone mad, and whom I suspect of being latent sensitives–by telling them, "You're not crazy, you've just begun to realize the potential of the human mind" As is my habit, on the night of the seance I arrived at Barrington House about twenty minutes early. Kathy came shortly after. While waiting for the others, we chatted. Perhaps to impress me with her knowledge of the subject (which may have been greater I suspected at the time) she related some of her earlier experiences with ghosts: in particular, spirit communications that took place before she came to work at Barrington. Admittedly, I do seem to find latent mediums everywhere I look. But when an active participant in a psychological experiment brings that kind of background into the seance room, the results are fairly predictable. Whether her subconscious mind originated the psychokinetic displays we witnessed or was merely the tool of an outside agency I was never privileged to discover.

Chapter 11: An Exorcism Made to Order

"Toward the unclean heart I mercilessly point my sword."
(Archangel Michael)

Mind-curists argue that to parapsychologists like me the line between the overly imaginative and the clinically mad is hardly noticeable. Yet many of my investigations are clearly absent of exaggeration, fantasy, and the ramblings of a disturbed mind. I ask that you judge this one for yourself. On Sunday, April 8, 1989, I got a call from former client, Lauren Kane: "I've had another *occurrence* here!" she said, sourly. "Two years ago ...you came to my house? ...Remember?"

Andrew: "Yes, of course, Mrs. Kane."

Lauren: "Well, my problem, or whatever you want to call it, is back! Nothing whatsoever has happened since you were here until yesterday ... more particularly, the night before is when I had the dream. Let me start with the dream, if that's what it was.

"I was in an open field. Straight ahead was an imposing figure, a man on horseback. He was pointing a burning sword at me...aiming it at my heart. Even when his horse, a huge white stallion, reared, his weapon was still directed toward me. I knew it was a dream, of course; at any rate, a fragment of me knew it. I am not prone to taking nightmares seriously, but this was no ordinary reverie. It was a truly traumatic experience!

"I woke up in a sweat. I cannot say when I first notice the grandfather's clock had stopped ... but it did at one thirty-nine. It quit at one thirty-nine before. Remember? This time there was *blood* smeared on the door in the shape of a cross." I hadn't expected to hear from Mrs. Kane again. Don't ask me how but I'd chased, or I assumed I had chased her disturbance merely by showing up the first time. As she spoke, I thought of that visit:

Andrew: (From the tapes) "April 7, 1989 ... it's 7:40 p.m. I am in the home of Lauren Kane. Mrs. Kane's daughters, Rose, Rochelle, and Chicky are here, as is Ed, Rose's husband. Rochelle's young friend, Ron, is also sitting in. (Names of family members and friends have been disguised according to their request.) "Mrs. Kane, fifty-three, is divorced. Her two unmarried daughters, Rochelle and Chicky, live with her. She has told us about some unusual things that have happened to her during the past several years ... both here and elsewhere. If you would, I'd like you to tell us for the record what's been happening here."

Lauren: "Well, the first time was two years ago, April fifth. The house that I owned burned. I had fire chiefs and the arson squad in and out for over a week. They never could determine the original cause of the fire. The house was essentially destroyed inside due to a break in a gas line. What happened was ... there was a small hot fire that started in the basement, maybe twelve feet from the furnace in an area where there was *nothing* that could have started a fire."

"As an afterthought, when we had the house built we put in a gas line for the fireplace. Instead of putting it up high, they looped it under the other pipes. A copper support for the line was directly above the flames when the fire started. When it broke, the pipe came apart and gas started to fill the whole downstairs area. The basement consisted of the furnace room, a laundry, a rec room, three bedrooms, and two bathrooms."

"At any rate, it was precisely nine-thirty. I was taking my younger daughter (she nodded toward Chicky) to the doctor's. And as we walked out the door it created a draft that was enough to ignite the gas that had filled the downstairs area. There was an immediate explosion. They never could find out *what* started the original fire. It wasn't a 'flashover,' you know, where everything catches fire at once; but the house was uninhabitable afterward. "So I would say that perhaps *that* would be the beginning of the, whatever ... the strange experiences. At that time, I was completely mystified. Of course, they were too as to what started it. If I'd been looking for gain or was a different sort of person they might have thought I started it. They asked my permission to bring the arson squad in, and I agreed. "I really didn't know what happened. In fact, when you're trying to sort things out you think, 'Who do I have as an enemy that might have wanted to start a fire in my house?' But there was no reason for it."

Andrew: "That was nine-thirty in the morning?"

Lauren: "Right."

Andrew: "You and your daughter escaped without injury. Your maid, you said she just barely got out the back door?"

Lauren: "She was in the kitchen polishing silver and managed to get out the back door before the fire spread upstairs. She rushed across to the neighbor's, which is a fair piece, and before she got there the police were on their way. I had a burglar alarm system in the house. The fire burned through the telephone wires which set off the alarm. As the patrolman was driving up he radioed back that it was a fire, not a break-in. So the fire engines were there in minutes."

Andrew: "You were lucky to escape without injury."

Lauren: "Well, it would have been terrible if it had happened while we were asleep. I think we'd have been dead." (The "psychic regulator" was on the job.)

Lauren: "Let's get off the fire! It's boring. That was 1987. It was a traumatic experience ...I know that. I didn't really associate (it) with anything ... anything out of the ordinary, or whatever you want to call it, until January of this year. A finger ring, a gift from my mother, had disappeared after the fire. It turned up mysteriously in January. I was greatly relieved and mystified, too, by its sudden reappearance. It turning up like that really bothered me. I could do nothing but stand there and stare at it."

Rochelle: "You took it out and looked at it once a day, as I recall."

Lauren: "I did! I looked to make sure it was still there ... that it wasn't going to vanish again. It was given to me by my *mother* after all."

Rochelle: "You even told the guy, the developer who bought your first house, to look for it in the wreckage."

Lauren: "That's true. I certainly did. It is very dear to me. Well, like I said, this is the first thing that threw me off."

Lauren: "The next occurrence that I can think of was sometime in February this year. (Speaking to Chicky) I'm talking now about the potholder. "I was smoking, like I am now, and smelled nothing. And Chicky said, 'Mother, don't you smell something burning?' And I said, 'No, I do not!' 'Well, I do, she said. And she went into the kitchen, and there was a potholder sitting on the stove, burning. I picked it up with a fork and threw it in the sink."

Ed: "The potholder was not near the pilot light, was it?"

Lauren: "It was on top of it."

Rose: "You've got to *see* this stove. There's like this grill thing with a metal cover over it."

Andrew: "Was it burning ... smoldering, or actually aflame?"

Lauren: "It was *ablaze."*

(Upon inspection I found insufficient heat radiating through the top of the stove to account for the incident.)

Lauren: "On the twenty-third of February ('89), I was next door playing bridge when Chicky came home-around a quarter to four. The house was locked tight. She came in here and the grandfathers clock door, which we always keep locked, was standing wide open. The hands were on one thirty-nine. Now this is a wind-up clock, not electric. Then I learned from the police that the security system had gone off at one thirty-nine earlier that afternoon. When they came they checked around the house but didn't find anything out of the ordinary. There was nothing missing in the house. The only thing amiss was the clock door. I even got the *chief* of police over here, and he looked at me like I'm out of my mind. I know he thought he was talking to some crazy old bag.

"Anyway, I started getting a funny feeling about the whole thing ... and at that time I had two detectors in this house: one gas and one smoke. I couldn't get them to go off, so I had them checked I think you'll agree, I had a good reason to be jumpy about now. They were both turned way down ... their sensitivity was way down. I didn't even know that you could turn them up and down. "Okay. Next occurrence. On the twenty-first of March, I'm in here by myself watching television. At ten p.m. there's a knock on the front door. I ask, 'Who's there?' There's no answer. And I open the door, and there is nobody there. And I say, 'Well, it's the wind.' But it didn't sound like that. It sounded like (she rapped three times on the coffee table). "So I go back to watch television, and at ten-thirty ... there it is again! I don't open the door this time. I turn on the outside lights, and I yell out. There's no answer. I peer through the drapes; I see *nobody.* Same thing happens at eleven o'clock. At eleven-thirty the doorbell rings: DING-DONG! DING-DONG! or, however it goes. Nothing!"

Rose: "At twelve o'clock, you're on the phone to me."

Lauren: "At a quarter to twelve, Rochelle comes home. And I say, 'Rochelle, do you think maybe I'm losing my mind?' I was very serious about this! Not that I'm nutty, but you can imagine an awful lot of things. Then, at precisely midnight, the doorbell rings again. And I say, 'Let's get out of here.' But, of course, we go nowhere. "And I'm in Rochelle's bedroom, which has a window facing the walk so you can see the front door. And *then* I call my daughter Rose. I'm sitting there on the bed looking out the window. I don't see how anybody could have approached the front door without me seeing them. And at twelve-thirty the doorbell rings again. Rose heard it over the phone. And no one is there."

Ed: "You heard it, dear?"

Rose: "She had the phone off in her bedroom and the one off she was talking on. And I could hear the doorbell ring."

Rochelle: "I heard it the last time it rang."

Ed: "And that's when you both screamed at your mother, 'Don't answer that thing!'"

Andrew: "You have pets."

Lauren: "Right. Three cats and a dog ... and they've never evidenced any alarm or anything. Now when the doorbell rang the first time, my dog, who is not much of a watchdog, but he did get up and sort of wander out into the hall. But he didn't bark. He didn't ruffle his hair. He didn't do *anything.* That's another reason I thought maybe I was going nutty. "The only other thing I've got to say is that on the twenty-sixth of March ... it was a Saturday night, I came home about eleven-thirty. I walked into the house, which first of all was very hot. I walked back to the bathroom and it was freezing cold in there. I sat down on the john and it was like sitting on a block of ice. I got out of there fast; and the hallway was still too warm. I stepped back into the bathroom...once, I think... and it was still cold. Well, I didn't go back."

Ed: "You didn't go behind a tree or something, did you?" (Laughter)

Ron: (Rochelle's friend) "Can I interrupt? I just remembered something. The night that I brought Rochelle home ... when you had the door knocks and the doorbell ringing ... when she went to the front door, I thought I heard something out back. It scared the beans out of me so I brought the .45 with me. (Shrugging his shoulders toward his frowning listeners) Well, I didn't know what to expect!

"I went round in back, trying to be very quiet to see if I could see anybody. And I went down the steps (the outside steps to the back yard), and I looked in the window. I didn't say anything about it before, but for an instant there was a light in the back room; and, when I looked in, the light went off."

Ed: "Were you down there, mother Kane? Was Chicky?"

Lauren: "No, I was not! Chicky was in the Bahamas; I was upstairs talking to Rochelle."

Ron: "I looked in the window and saw a light for an instant in the very back part of the furnace room."

Rochelle: "That's a single light bulb on a chain."

Ron: "When I came in the house, the first thing I did was to check downstairs. I didn't find anything, and the only thing I heard was the ice maker kick on."

Andrew: "How would you describe this light? Kind of like a light bulb?"

Ron: "No. Not a light bulb. I didn't see a light source. For an instant the furnace room was lit as bright as day. I likened it to the image of a flashbulb because it was white light. It was so impressive to me that I had chills."

Andrew: "Let me ask you one more question, because maybe we're getting into familiar territory here. I don't want to plant any ideas in your head, so if I'm wrong correct me. You know how a flare looks the moment it explodes or the way sparklers look? Can you compare it to that?"

Ron: "It was bright enough to light that back room. And the instant that I looked, it went out ... like maybe my looking in the window was a signal for it to go out."

Andrew: "Mrs. Kane, has your family or your husband's family a history of strange occurrences, spontaneous fires, or psychic events?"

Lauren: "I cannot say for my ex-husband's family. I do not *think so.* As for my family ... definitely not!"

Andrew: "And you never had anything of this nature, except for the fire at the other house?"

Lauren: "Never!"

Rochelle: "What about the time your adding machine started running itself?"

Lauren: "Honey, that was due to lightning or something or other."

Rochelle: "No more than your doors blowing open and slamming shut, and your windows opening by themselves."

Lauren: "I don't know. The only thing I can say is that to me this is confusing. If it is something ... what do you call it? Supernatural? Oh-cult? Is that the correct pronunciation? Whatever ...I can't understand two things: Why doesn't it bother my animals, and why at no time through any of this have I felt any sense of personal fear ... none whatsoever? "I am essentially a very logical woman. I've never believed in spirits or anything like that. That's the kind of person I am: too hardheaded, I guess."

And that's all there was. Except for the mysteriously generated fires, there wasn't much to ponder. The ring, turning up the way it did, could be attributed to the chaos following the explosion: it was probably there in her jewelry box all the time. The open clock door, the alarm having been triggered, and the time coincidences were probably just that--coincidences. Security systems are temperamental gadgets, as anyone who's been awakened by one going off accidentally will attest. The knocking and bell ringing were limited to just the one episode; the cold bathroom, a subjective experience at best. There were the malfunctioning detectors, and the flash seen near the furnace (the fire in the other house had started near the furnace). Yet neither were overt signs of a haunting in progress. Parapsychologists label arsonous psychic phenomena, "Incendiary Poltergeists." Some of these mind-projected pyromaniacs do not stop until their victim's home has been totaled: a tidbit we kept to ourselves.

In the spring of 1989, Lauren was still Lauren: calm but curious. "I called," she said, "because the whole situation throws me. You put a stop to it before (in truth, it had stopped on its own). Do you have any notions about the dream or the blood on the clock door? I'm not going nutty again, am I?' I told her I honestly did not know what either signified--if, indeed, they signified anything.

As I listened to her raspy voice, I recalled that Lauren Kane was the toughest ninety-pounder I'd ever come across. The matriarch of her family, she is a wiry, strong-willed, opinionated old bird: the kind of woman who does not suffer fools easily. During the period of our first visit, our client, together with her three brothers, operated a heavy equipment leasing business. She was not only educated, she was "street smart." To her, such inane explanations as Fire Poltergeists and door-knocking mind-creations were not only poor substitutes for an in the flesh villain, they were sheer and utter nonsense.

Not satisfied with the efforts of the police or fire departments, Lauren was looking for answers in less orthodox places. She was not prepared to accept those as unconventional as ours. But, then, we were still two years away from all hell breaking loose.

Psychic Attack

Two months later, Lauren's daughter called me at my office:

Rochelle: "I'm still living at home, Mr. Nichols. Chicky's in school in California. She'll be home for the summer in a couple of days, but for now it's just my mother and me. Wednesday night there was something that looked like tacky blood all over the wallpaper in my sister's bathroom. I mean it was streaming down the walls. Looked like it'd been spread on with a paintbrush."

Andrew: "Is that the bathroom your mom said was so cold?"

Rochelle: "As I recall, it is. We tried to clean it off. Most of it slid come off; the covering is vinyl so we could wash it. This morning we got up to the same mess. I have no idea what it is, but it definitely looks like dark-red blood with a kind of watery puss in it."

Andrew: "Did you see the cross on your grandfather's clock? Is it the same kind of substance?"

Rochelle: "I don't know. Mother cleaned it up before I got to see it."

Two days later, on Sunday, June 3, our "Poltergeist" had evolved into something more interesting. I was in the office catching up on a few things when

Lauren called. Chicky was home and had seen what she described as a zigzag thing her first night back. Lauren put her on the phone:

Andrew: "What's a zigzag thing, Chicky?"

Chicky: "Not thing ... THINGS! Last night, around two-thirty, I heard scratching sounds coming from the chifforobe. I only was asleep for a few minutes when it woke me. Tabby gets trapped in there sometimes when I close the door on her. I got up in the dark, went over and opened the door so she could get out, then flopped back in bed. "Ten, maybe fifteen minutes later, I heard it again. This time it sounded like something was clawing on the wall behind me. I turned on the light but couldn't see a thing at first. Then I saw them! Three of them about twelve inches across ... hovering in midair. Zigzag balls of hair; purplish-brown; no heads, just claws hanging down. They were tucked under, the way a bird holds its talons, but you could still see them. I said, *WHAT AM I ON?* I tried to scream, but nothing came out. Then I did and they went up in a puff of smoke."

Andrew: "What are you on, Chicky? Seriously!"

Chicky: "No, no. I don't do drugs (spoken matter-of-factly).

Lauren: (Grabbing the receiver) "She let out a yell that could have awakened the dead. It startled me awake, and I sleep like the dead. I told her it was a dream, but I don't think she fell back to sleep. Did you? The grandfather's clock-which hasn't worked since I called you in April ... started going strong again last night. And the dream's back. The horseback rider is still aiming his spear or sword, or whatever, at me."

The next day, Monday, late afternoon, I heard from Lauren again. She was in a dreadful state: "I'm in trouble ... bad trouble, I think. I can take a lot of punishment. I have all my life, but not this." Around 4:00 p.m., while in the process of rearranging her living room bookcase, Mrs. Kane suffered three parallel lacerations. Each wound, a jagged scratch, was about five inches long and extended nearly the length of her frail forearm. Needless to say, this was no ordinary Poltergeist...

Lauren: "I didn't notice it right away. I was wearing a long-sleeve blouse buttoned at the wrist. Suddenly my left arm felt warm and damp. When I looked, there were streaks of blood oozing through my sleeve and dripping down my hand onto the carpet. I unbuttoned my cuff, wiped some of it away and saw the scratches: claw marks like one of the cats had gotten me. But my cats were all neutered years ago, so I don't know. Her voice stiffened: "I cannot say how they occurred. But I will say this ... There is absolutely no cut or tear in the sleeve itself. I was deeply scarred without my blouse being damaged, whatsoever. I know that is impossible, and I'm very much troubled by it."

The wounds bled profusely but not long. When they stopped she washed her arm, applied an antiseptic and a gauze dressing. Later, to the girls she passed the bandage off as nothing serious: "Tabby caught me. That's all."

She woke up in a pool of blood the next morning. The dressing was drenched and when removed revealed three new gashes, each about two inches in length: "They're at right angles to the first three ... parallel to each another. My arm looks like a grotesque game of tic-tac-toe. I do not, I cannot fathom how I could have been cut when the bandage remained intact. Like the sleeve of my blouse, as far as I can tell no part of it was penetrated. It is completely whole with no tears ... impossible, but true! "One thing, those from yesterday are nearly healed today. They're just long, light-brown scars. They were on their way to being healed when I talked to you (which explained why she chose the words "I was deeply scarred" instead of cut). "That's impossible, too. It's all impossible, isn't it?" Not yet panic-stricken, but close to it, she called me: "There is absolutely no way we're going to spend another minute in this house!" she burst out. "We're leaving!" Thinking back to the fire and how lucky they were to escape with their lives, who could blame her. At six-thirty she called me again. She'd grabbed Rochelle and Chicky and headed for a motel. "That's where we'll be until further notice," she announced.

Standard procedure requires that I bolster the confidence of the haunted by assuring them that nobody has ever been harmed by a ghost. It's a psychological tool that works most of the time. It not only quiets their immediate fears, it allows me to put them off till the disturbance "matures" (a euphemism for going away on its own). Lauren Kane would be put off no more: "I'm sick to death of this! I want you at my house, Thursday evening, seven o'clock sharp! Don't be late. We'll be in the driveway, waiting."

But that was two days away. In the meantime, they had to live out of suitcases. Rochelle and Chicky adjusted quickly, pouncing on the queen-size bed facing the TV. Their mother took the one nearest the door, grumbling all the while about the dingy bedclothes and lumpy mattress. Lauren was uncomfortable but less apprehensive now that they'd escaped the house. She was frightened for herself; that she couldn't deny... But it was the safety of her girls that mattered most. Days later she told me, "I'm an old broad. I've had my life ... my daughters haven't. I thought the house was the key, so I wanted them out. Even if I had to give the damn place away, we were out of there! We were plain lucky the first time. Who knew what was in store for us this time round?"

Thursday night arrived. On the way to the house I considered the possibilities. A séance was risky. Mrs. Kane was undergoing some kind of assault. It was too early to tell whether it was from within or without (though I was nearly certain it was the former). In either case a seance might heighten the effects--effects that were already out of hand. The roadside hotel turned out to be no haven. In the middle of a fitful sleep Chicky shot awake. In a clogged voice she relived the experience:

"Talk about your *'Motel Hell!'* I couldn't believe it. This screechy sound was going up and down the wall like nails on a blackboard. I wanted it to be a dream, but I knew it wasn't." Rochelle, awake now, looked across at her sister silhouetted against the window. As she did, Chicky reached for the lamp on the table between the beds. "I never saw or heard anything ... not really," Rochelle admitted "One of those things was hovering near our mother," said Chicky. "Nobody, but nobody can scream like my little girl. She could shatter lead with that mouth! "

Lauren hadn't seen it either. But then she was busy with her arm. Jolted awake, she'd watched in silence as a dark spot appeared then spread the length of her bandaged arm: Lauren: "Again, the dressing was completely intact. We were up the rest of the night. I checked us out at five that morning. Is there no place where we'll be safe from this thing?"

Andrew: "You appear to be experiencing a psychic attack, Mrs. Kane." (Her face contorted into a *No shit, Sherlock* grimace.) "Whatever its source it is not, it seems, linked to your house ... the way a true haunting would be. You'll be just as safe here as anywhere." "These things are probably attached to one of you: either you or one of your daughters. They're a little old to be projecting Poltergeist phenomena, if that's what it is. But it's not unheard of."

Lauren: "Are you telling me one of us is *possessed* in the Catholic sense?"

Andrew: "No, no. It's more a question of being *oppressed.* "

Lauren: "We haven't discussed it, but I'm not religious. I'm not the least bit superstitious. I don't believe in it any more than I believe in ghosts. I did go to church when my mother died last October, but that was the first time since I got married. In your opinion, do you think that's causing these problems ... my lack of religion, I mean?"

Andrew: "Unh-unh. Now let's examine your arm." With Rochelle's help, Lauren gently peeled the dressing. It was a sight beyond description. Her forearm had been through a meat grinder: a dozen crisscrossed lines of scar tissue running every which way. The topside from wrist to elbow was transversed by jagged wounds in various stages of healing.

Andrew: "Are you in pain when you're cut? Do they hurt?"

Lauren: "It's more unsettling than painful."

Andrew: "Do you have a flash camera? I want you to keep a photo diary of these attacks; take snapshots of your arm as they occur. For now, though, I can *protect you* from them. Do you have a tube of lipstick I can borrow. Any kind will do, even an old one."

The girls helped their mother re-bandage her arm. Using the lipstick, I drew four "mystical symbols" on it: ancient protective sigils found in medieval magical texts such as the *Key of Solomon.* Then I used clear adhesive tape to seal the ends--just below the elbow and above the wrist. During the weeks that followed, Lauren was to suffer forty-eight additional lacerations to her left forearm. A few broke the skin where the adhesive tape had come loose. The rest were inflicted when she showered and when she changed the dressing. Not once, however, was she cut in an area protected by the symbol laden bandage.

So why the "mumbo jumbo?" The only thing magical about the protective symbols I drew on Laurens bandage is the fact that they'll probably work. I was sure that the woman was in such a pliant and suggestible mood they were bound to be effective. People are far more dynamically linked to their ghosts than they may think. It didn't matter what I drew; he could have painted a skull and crossbones. What mattered was that *she believed* these signs would protect her; and so they did. It was the first clue to the source of the attacks.

Logically speaking, there was no way her arm could be cut as long as it was securely covered. But how explain the numerous times she'd been clawed while it was? I hadn't stopped them completely, not yet; but limiting the attacks was in itself a minor miracle. On Friday evening, Lauren phoned me again. She'd endured four more scratches while showering. They started just below the elbow and curved in to form a single line ending about an inch above her wrist. A new pattern was taking shape. No longer a sadistic game of tic-tac-toe, the saw-like cuts now resembled an arrow. Just below the elbow, we could make out the feathered ends. The shaft continued down the outside of her forearm to the two large bones in her wrist; there, a jagged arrowhead was clearly visible. What it all meant was anyone's guess.

I re-bandaged the arm and drew the four symbols. I was about to leave when Chicky came into the foyer sobbing. She'd seen the hairy things hovering in her bedroom. I rushed in (not too quickly, I'll grant) but found nothing. I was still in her room when I spotted two native-looking ceremonial pieces that were to figure prominently in the investigation. Earlier that year (1979), Chicky and three of her classmates had gone to Kenya to visit the Maasai tribe. Although the local government had restricted tourists, a bit of Yankee ingenuity and a sizable bribe managed to get them into Maasailand and inside one of the tribal villages.

Snapshots of grinning villagers documented the story. Putting the photos back in their decorative box, Chicky handed the first of two prized possessions to me. It was a large, elaborately decorated neckpiece that, when placed over the head, covered the upper part of the body from just inside the shoulders down to the breastbone. The second was a brightly colored beaded necklace with a leather pouch attached. At first glance, both appeared to be run-of-the-mill tourist trinkets. Now, as I examined them closely, I could see they were finely crafted ceremonial pieces of some kind. Noticing my interest in the leather pouch, she said, "Scott, my friend, had a hard time talking this guy out of it. He wouldn't tell me what he paid, but I think it was like thirty dollars. That's a lot over there. I stashed them in my footlocker at the dorm and didn't open the pouch till I got home. There was a dried up, crumbly wad inside; I think it was dung. I dumped it, washed the pouch and hung it on the wall behind my bed." "When did you do that?" I wanted to know. "Last week ... the day I got home from school. Come to think of it, it was the night I saw the zigzag things."

One point about this case troubled me: the lack of an independent witness to the attacks. No one, not even Lauren herself, had seen the breaks in her skin as they occurred. It was conceivable that they were self-inflicted--without conscious volition, perhaps--but self-inflicted nonetheless. There was another possibility, a rather exotic one but worth considering. Lauren was a world traveler and world class collector--as evidenced by the hundreds of handicrafts that adorned (more accurately cluttered) her family room. Could she have contracted a virulent skin disease, or picked up a parasite in a foreign land--something that was causing these freakish wounds? It was an area beyond our knowledge, so we decided to delay further action until she got professional help (not that we had the slightest idea of how to proceed at this point). She needed medical, psychological, and theological help--the latter thrown in for good measure. Much to our surprise she agreed to apply for all three. On Sunday, June 10, she reported the appearance of four more gashes. Again they came while she showered:

Lauren: "I never see them when they open. It's like the movie Psycho: I catch sight of blood as it drips onto the shower floor, or as it begins to mix with the bath water and swirl around the drain.
"The man on horseback appeared last night, pointing his flaming sword at my heart. He's considerably closer now. I would say he's wearing a suit of armor. His determination is chilling ... like someone walking across my grave."

Monday was without incident. But Tuesday, June 12, ushered in a new mystery. At 12:30 a.m., while reading in bed, Lauren felt a cold breeze strike her face. She looked toward the window and saw a white, misty shape near the fireplace: "It was five feet tall and one to one and one-half feet wide. It took no human form or any other identifiable shape. It raised more than a few hackles, I can tell you. But I was strangely unafraid; I don't think it meant to harm me. It glowed from within and was visible for two or three minutes before gradually fading. There was a definite coldness in evidence all during the sighting. Once the vision or ghost, if that's what it was, was gone, I summoned the courage to walk to where it had stood. That spot was colder than it should have been ... similar I think to the coldness in the bathroom that time." "You know, Lauren, fear can trigger feelings of coldness," I explained. "That's why we shiver sometimes when we're scared ... to create body heat."

"Yes, but I felt the cold breeze before I saw the misty form. In any event, if that's what ghosts are about I have nothing to fear from them. I wasn't scared, only curious and slightly apprehensive.

"Later my knight invited himself into my dreams again. He was closer than ever. It's not the ghost or even all the blood that bothers me ... It's him! His sword is pointed precisely at my heart. "Tomorrow, in the afternoon, I see my M.D. Good luck to me. Right?"

Lauren was well-acquainted with her doctor's temperament. He'd never understand the symbols on the bandage (no more I imagine than would my physician or yours). She shoved the cast-like covering into her purse, then settled back in the waiting room chair. Five minutes later, a three inch gash opened on her forearm. It was a familiar scene. Her long sleeve blouse was buttoned at the wrist; there was no pain; and only the sensation of the polyester sticking to the wound, followed by the sight of stains alerted her to the attack: "Those sitting near me were aghast at all the blood! Strangely enough, I was elated. I believed that it would serve as tangible proof that something remarkable was happening to me. But my hard-ass doctor would have none of that!"

When I suggested she see her physician it was with the hope that, if he didn't recognize the problem, he'd refer her to someone who would--a dermatologist, perhaps. If a specialist found no medical reason for the wounds I presumed he would recommend a different course of action. Maybe even psychiatric help. I had presumed too much:

"He took one look at all the scar tissue and the still oozing cut and handed me a dictionary. Looking me straight in the eye, he said, 'You know the word inebriated? Look it up if you don't.' I started to shake and said, 'You know the word incompetence?' This is a man I've been going to for over twenty years; but no more!" I waited for her anger to subside, then forged ahead. "About your drinking ... Do you have a problem with it?" "Absolutely not! Do I drink some? Yes, of course. Who doesn't, and who wouldn't in my circumstances? Do I drink to extremes? *I do not!* At any rate, not to the point where I start sticking myself to see if I'll bleed. "My *doctor* ... and I use the term loosely ... told me it wasn't possible to get mysterious lacerations. Either I or someone else was deliberately causing them. Yes. That is the only logical explanation. I know that! But what does it prove? *Nothing!"* If Lauren Kane had a problem with booze I hadn't seen it. She was always in full control of herself when I was with her--mentally as well as physically. I marveled at her ability to stand up under the now almost daily attacks without resorting to *drugs,* much less drink.

At around ten on the night of June 13, Lauren called to say she'd suffered a five inch cut. No sign of her ethereal visitor, but the area was "ungodly" cold again. She would see her minister on the fourteenth. She was not thrilled at the thought, and told me so. I half-expected her to back out. The fourteenth came and no word from Lauren. I called her: "1 could've told you what Rector Aldrich (a pseudonym) would say. He doesn't believe in witchcraft and magic and things like that, and people who do are condemned to eternal damnation. So where does that leave you and me, bub? Where do I go from here?" Neither scientist nor priest had given solace. Psychiatrist Carl Jung took both professions to task for their shortcomings in this regard when he wrote: "Both doctor and clergyman stand before (man) with empty hands, if not what is worse, with empty words." Lauren Kane had been turned down when her need was the greatest. Now I was about to do the same. The case had simply evolved too rapidly. I couldn't help her until all normal avenues had been explored. And after that? I hadn't a clue how to approach her unique situation.

Then I remembered a local practitioner, a man named Pauling (a disguised name) who was a medical doctor and psychiatrist all rolled into one. I hoped he'd be willing to take on this head-strong woman. She protested but finally agreed to see him on the eighteenth. The night of the fifteenth "Sir Knight" was back, her dream more lucid than ever. His visor was lowered, yet she could see his piercing eyes through the diamond shaped slots. The tip of his sword, on a line with the helmet, was bent slightly downward on an arc that would eventually put it through her heart. Calmly, almost indifferent to that fate, she said, "Whether this is a dream or not ... which, I am no longer competent to judge ...I know it's only a matter of time before he overtakes me." All the while the attacks continued. A milestone was reached on the sixteenth. Rochelle was helping her mother with the dressing when she observed the spontaneous appearance of a four inch scratch. She described it as similar to a rough-edged paper cut: dry for a few seconds then bleeding copiously. Lauren was vindicated. Her arm had opened on its own in front of a witness. I was more comfortable with the idea that the wounds were not self-administered, yet still not totally convinced. Corroboration had come from her daughter and not from an impartial observer.

I was happy to hear she'd kept her appointment with Dr. Pauling; 1 wasn't sure that she would. His approach was the correct one--upbeat, positive: "For the first time I'm optimistic," she said with conviction. "I really think he's going to help me ... and speedily." Pauling, on the other hand, had his doubts. Because patience is not one of Lauren's virtues he had suggested I use hypnosis to stop the assault in its tracks. Something had to be done and fast or her confidence would vanish and she'd walk. As fate would have it, the woman is one of those rare individuals who cannot sustain trance. We guessed that her treatment would be lengthy. She was back in his office the next day for a conference. Her subconscious mind was inflicting the wounds owing to a deep-rooted guilt complex he told her. It was a condition that could be worked out in time with a combination of psychotherapy and reinforced suggestion. Meanwhile, his goal was to alter her habits: the excessive smoking and drinking. My charge was ill- prepared for the long haul: "He's telling me the scratches will stop in time. But first, the key, he tells me, is self-control of my bad habits ... which *he* finds excessive. My God, I'll be dead before then! I'll bleed to death, be run-through, or go insane before *he* helps me! Why can't the man just focus on first things first?"

They were classic signs of psychological resistance (opposition to orders that affected her lifestyle). A kind of panic came over her as she pleaded with me to find a way to terminate this whole business:

Andrew: "Your problems have been going on for years ... off and on. They're not going to disappear overnight. At least, give the man a chance!"

Lauren: "Do you know my business and social life have come to a standstill? Where can I go wearing that ridiculous bandage? How do I explain it; the stupid thing won't fit under my sleeve. I'm afraid to wear it and face the asinine questions; but I'm more afraid not to. Worst of all, I worry about my girls. What if it turns on them?"

Andrew: "I think if it meant them harm you'd have known it by now."

Lauren: "You think! Oh, that's comforting. (Calming down a notch) Whatever this thing is, I have come to believe that it is essentially evil. Something evil is persecuting me, for what reason I cannot say."

Lauren continued to see Dr. Pauling. I don't know what I'd have done if she hadn't. From the eighteenth to the twenty-second of June she underwent a total of eleven lacerations. No longer was there a recognizable pattern to them. Photographs are appalling. In one, her forearm is seen against a white background: a sheet draped over a couch or chair. Thick blood lies in small puddles along the line of a recent scratch. In some places it rolls down the arm, then drips off. Cuts in the process of healing have turned her skin to a bright, almost glowing reddish-brown. The "feathered" ends of prior attacks can still be seen. In another shot the arm is raised, bent at the elbow and pressed against the left side of her face. Seven trails of blood flow around to the underside, spill over her biceps and fall like raindrops. It is a gruesome, gruesome sight.

On Saturday, June 23, for only the second time, a witness saw an attack in progress. At 3:20 p.m., while her maid helped with the bandage, Lauren sustained a four inch gash. The incident was confirmed over the telephone. She awoke with a start the next morning. Still aimed at her heart, the blade of the flaming sword was now no more than twenty feet away. The knight leaned forward in his saddle; blue-white vapor escaped from the stallion's nostrils as the powerful animal galloped toward her: "I was paralyzed with fear. He moves in slow motion, or I would be shish kebab by now."

A series of uncanny events occurred the night of the twenty-fourth. While Lauren was in the shower, the front, back, and garage doors opened simultaneously setting off the alarm. A police cruiser was on the scene in minutes. Just as it rolled onto the driveway, she experienced the most severe laceration yet. It extended from elbow to wrist.

Not taking the time to put on a fresh bandage, she hurriedly dressed to go downstairs. Just after the officer left and before her arm could be protected, another jagged break appeared parallel to the first. On the twenty-fifth she saw Dr. Pauling again: "I'm somewhat happier with him now than I have been. Did you mention the disparaging things I said the other day? (I hadn't.) Well, like I said, I feel better about it now. This was the first time he talked to me directly about my arm. He thinks I'm doing it to myself. .. through my own mental powers." (She didn't elaborate.) She returned to his office the next day. As soon as he started in on her bad habits--back in the doghouse he went! What little patience there was had been worn thin. She wasn't up to another lecture.

Breakthrough! Wednesday, the twenty-seventh. On my way home from the office I stopped by the house. At approximately 6:00 p.m., in the middle of examining Lauren's arm, the telephone rang. She turned, took a couple of steps and picked it up. Instead of "hello," she shouted, "oh!" and dropped the receiver. I was there in an instant. I grabbed her wrist with one hand and cradled her elbow in the other. A cut was making its way down her forearm. It was seven inches long by the time it reached her hand. "It looks as though someone with rough nails got me, doesn't it?" she muttered. It resembled a paper cut more, I thought. Anyone who handles letterheads, envelopes, even pages of a book knows what these painful slices are like: bloodless for a few seconds, then droplets bead along the line and it begins in earnest. Thirty minutes later as I was leaving for home the wound had already closed and begun to heal.

No police matron could have frisked the poor woman more thoroughly than I. There was nothing. Nothing that could have been used in the attack: no abrasive or keen edge. I examined her outer and, I'm embarrassed to say, her inner garments; the telephone stand; the shag carpet she stood on. I even checked her fingernails, bitten down to the quick. I concluded that it would've been difficult for Lauren to have hidden a sharp object on her person, in her clothes, or nearby. I was astonished by what I'd seen. Still, there was this uneasy feeling. Her reactions weren't normal. My client was bewildered, yet strangely unperturbed. Not a tear was shed. There was no unspoken terror in her voice; no alarm in her actions. She was either the bravest woman I'd ever come across; numbed into silence by the repeated attacks; or else, adept at keeping a tight lid on her emotions.

On Thursday, June 28, Lauren got a reprieve; but the attacks were back with a vengeance on the twenty-ninth. From that day through Wednesday, July 4, she was cut five times. On the fourth, I got a phone call from a desperate woman: "I am willing to undergo *anything* to stop this damnable thing," she begged. "No matter the danger to me personally." Again, I asked that she be patient with her doctor's methods. She shot back, "I'm sick of that hooey! It's too much to suffer through all this and give up my only crutches. I'll *bleed* to death before I can do that." On the night of the fourth--and again on the fifth, and sixth--Lauren dreamed of her nemesis. His sword was threateningly near.

Sometimes the obvious completely eludes us: not a method to stop Lauren's attacker--that was far from obvious. There was something less complicated that she could do to protect herself: a step that never occurred to me. A colleague of mine thought of it. She wondered why, instead of removing her protective covering every time she bathed, Lauren couldn't wear a long, oversized rubber glove over her bandaged arm. Also, why she didn't change dressings once every two or three days instead of every day, especially when the attacks were less frequent. These few simple measures reduced her discomfort significantly.

Somehow my client got the idea that she needed an exorcism. She mentioned it to me for the first time on the fifth of July: "That's what I want," she implored. "It's the only way to rid me of this evil." Strong skepticism often gives way to even stronger belief. A hard-boiled disbeliever for so many years, Lauren Kane had become an overnight convert to credulity. She even volunteered a priest who would be more than happy to perform the ceremony: "Chicky's best friend ... her grandmother's friend's brother was an exorcist: Father Anthony (a pseudonym), a Jesuit. He was at Alexian Brothers' Hospital with Father Bowdern back in 1949, when they did the real exorcism (referring to the much publicized case on which William Peter Blatty based his bestseller, The Exorcist)." "I think that'll take some doing, Lauren," I countered. "Before an exorcism can be performed it must be sanctioned by the bishop of the diocese ... and then only after a complete investigation is made." "Oh. I do not, I think, have time for a complete investigation." "It's not the lack of time that worries me," I told her. "Frankly, I doubt that you can pass an investigation--especially the medical portion. But, even if you can, your symptoms are not--strictly speaking--those of the possessed, but rather those of a person suffering Oppression or 'Obsession.'"

Freud was the first to use the term Obsession instead of the more dramatic and religiously oriented word Possession. But Freud was thinking in terms of a psychological malady. This merciless persecutor, whatever its source, was physically assaulting her from without--not from within; or so it seemed. In this country, exorcism is still practiced, albeit infrequently, by the Roman Catholic, the Episcopal, and even less often by the Lutheran Church and the Jewish religion. The ceremony is also part of Anglican, Greek Orthodox, and some Protestant fundamentalist denominations here and in Europe. The evidence required by most as proof of diabolic influence consists of four elements: speaking in tongues, knowledge of secret or distant facts and events, the presence of an alien intelligence, and superhuman strength. None of these fit Lauren Kane's condition. Subtle exorcisms: a pinch of salt tossed over the shoulder; a "God bless you" after a sneeze--these go on all the time. They're merely ways to ask for protection from evil. The solemn side of diabolical influence has to do with possession: losing control to an objective intelligence, and obsession: being tormented by devilish forces.

Clerics and ghost hunters everywhere hear from those who believe Satan is tormenting them. Many of these "put upon" people develop contagious obsession-possession (i.e., psychosomatic symptoms), or they imagine these conditions in family members or friends thanks largely to exposure to dogmatic literature and accounts of ritual exorcisms in the media. Good counseling is often the answer. Blessing the house, an informal rite performed by the clergy, is another. Exorcism is the last assumption organized religion makes. It was certainly far down on the list of mine.

"Forget the religious ceremony," she said. "I want you to do it." "You mean a layman's exorcism?" "Whatever. I'm comfortable with it. You got rid of Poltergeists before." When I passed on the dubious suggestion to Dr. Pauling he astonished me by giving his approval. There is no taped record of our conversation, but my notes indicate it went something like this:

Andrew: "Mrs. Kane is desperate. She's asked that I perform a layman's exorcism to expel this thing. I'm concerned that further damage might result if we do. She tells me she doesn't believe any more damage can be done to her."

Pauling: "She's in bad physical condition, but good, I would say, mentally (the opposite of my diagnosis). If she insists on it ... go ahead. It probably won't do any harm. And who knows? It might just shock her back to reality." He assured me that Lauren was not suffering from disease, a parasitic invasion, or any other physiological problem that could account for the scratches. Although he saw no advantage in the ceremony, neither did he think it could cause her additional distress--that is, if it was conducted properly. Surprisingly, there are medical professionals who, although adamant about the reality of obsession-possession, are sympathetic to the use of exorcism. Hypnotherapists in particular are aware of the beneficial use of suggestion during controlled and structured rituals.

I had already set the date: Thursday, July 12, 1979. "Would the good doctor be joining us?" Was I kidding? Lauren managed to get holy water from Father Anthony. I told her to sprinkle it around the house: a precaution not entirely successful. On the seventh, she sustained a cut while changing the bandage; two more on the ninth.

Wednesday the eleventh--the night before my scheduled visit--her nightmare had reached its pinnacle: "I may *never* sleep again ... not until it's over. Tuesday, he was a hair's-breadth, or should I say a horse's breath, away. Last night, his blade was pressed sharply against my flesh ... ready to pierce it. The shaft was aflame; the point, indescribably cold and painful." By the eve of Friday the thirteenth, Lauren Kane was in the final phase of her ordeal. She refused to go on unless her life changed; she was resigned to whatever that change might bring: "I feel like I'm in death's grip," she observed. "I am like a patient waiting for an exploratory operation. Even if I go beyond the brink, it would be better than living *this* way." I felt I had no choice. I had to go on with it.

One of the first recorded incidents of psychic attack occurred in 1762 when two young girls in England were wounded by some invisible force. *(A Narrative of Some Extraordinary Things,* H. Durbin, Bristol, 1800.) In 1935, a Cincinnati housewife was badly cut according to a local newspaper. Thurston mentions such attacks in *Ghosts And Poltergeists,* (Burns and Oats, London, 1953), as does author A.R.G. Owen in *Can We Explain The Poltergeist?* (Helix Press, New York, 1964). In 1990, the Staff at Llewellyn Publications put out a booklet called, *The Truth About Psychic Self-Defense:* a manual on how to protect yourself from psychic attack and energy drain.

One of my favorite accounts appeared in William G. Roll's, *The Poltergeist* (Doubleday, 1972). Roll wrote about the mysterious bite marks on the arms of three women in Indianapolis, Indiana in 1962. There are countless others, but only a few have been properly documented. I didn't put much stock in it, but Mrs. Kane had come to believe that the five dozen slices out of her arm were the result of an onslaught of demonic forces. I was obliged to treat the problem accordingly. The idea that the devil was after her could be traced to an event that took place the day before the attacks began. She'd been assaulted initially on the fourth of June. The previous day, she and Chicky had seen the return engagement of *The Exorcist* the movie based on Peter Blatty's best selling novel. Neither had been exposed to the story before.

The Exorcist is a movie that turned more heads than Linda Blair's. My readers may not remember the hype that followed the first showings, but it was said that some in the audience (no doubt the emotionally unstable) rolled in the aisles, convulsed and fainted. Clergymen across the country accused Blatty of "setting the church back 400 years." Psychologists suggested he was responsible for triggering all the pseudo-possessions that cropped up in cities where the film was shown. Those affected were highly susceptible to suggestion. Lauren Kane, according to Dr. Pauling, could be numbered among that group.

Blatty asserted the story came from a secret diary kept by Father William Bowdern, pastor of St. Francis Xavier (St. Louis University) College Church, and the man charged by Archbishop (later Cardinal) Joseph E. Ritter with performing the now world famous exorcism. Somehow the author got hold of a copy. An article appearing at the time of Father Bowdern's death (in the April 26,1983, *New York Times*) reported that, in 1949, the Jesuit exorcized--with the help of several other priests a fourteen-year old boy from Mt. Ranier, a suburb of Washington, D.C. A study of this case became the basis of the fictionalized account in *The Exorcist.*

What has any of this to do with Lauren Kane? Possibly nothing. But I find it interesting that Father Bowdern's account, according to those who have read it, begins with the boy hearing scratching sounds coming from the walls. In addition to other vile acts committed against him, he was repeatedly clawed through clothing and through sheets and bedding by something invisible that left no mark on them. On one occasion, zigzag scratches were seen on his body in the form of an arrow.

The enigma deepens. In the Blatty version, the culprit was Pazuzu, a particularly ugly Babylonian demon pictured as having talon feet and ragged wings. Our client avowed she knew nothing of Pazuzu or of the wounds suffered by his victim. Except for a statue unearthed in Iraq, not much about the "little devil" was portrayed in the 1973 film, and even less was made of the scratches (although the word "help" was spelled out in welts on actress Linda Blair's stomach). Green vomit, a swiveling head, and a levitating teenager are bound to create more ticket sales than an occasional nick or two on the arm. Clearly, it is difficult to ascribe Lauren's ordeal to suggestion brought about by what she saw and above all by what she *didn't* see on the movie screen. Yet the sudden attacks began the next day! It is a staggering coincidence.

In 1949, all attempts to treat the boy medically and psychiatrically were unsuccessful. Not until the exorcism was completed, an achievement that took no less than twelve weeks, did the brutal assaults stop. According to a psychiatrist who reviewed the case, the lad's restoration to health demonstrated something I know well: that what is induced by suggestion can be cured by suggestion.

Lauren's wounds, like the wounds of those manifesting the well-known hysterical symptoms of stigmata, were probably externalized repressed emotions finding a safety outlet. In stigmata, nuns, often novices, become the victims of the devil and suffer the passions of Christ. And, like the afflicted novice, only a religious (or in my case, a quasi-religious) ceremony will remove the evil influence. Lauren Kane was no nun. Even so, if the movie had somehow triggered the attacks, a properly staged ceremony had a good chance of interrupting them.

From the shoulders down, artist's renditions of Pazuzu--though considerably larger--bore some resemblance to Chicky Kane's headless visions. Did these "fuzzballs" play a role in the mystery? I spent a Saturday afternoon in the public library researching the Maasai. When I was through I still knew practically nothing about them, except that Chicky's large beaded ornament was typical of their handicraft. Her necklace was a different matter. In a booklet prepared by the government of Kenya, a warning was issued to the native population about practices of fraudulent "Doctors of the Arts." (They don't call them "witch doctors" over there.) In part it read: "There are two types of doctors in Maasailand, namely the genuine ones whose methods and diagnoses are clear and obvious, and the tricksters whose only interest is to enrich themselves by charging high fees. "Sometimes when a patient goes to one of these (tricksters), he's asked to go outside and fetch cow dung (the Maasai are a cattle raising tribe). Then he is asked what and where his complaints are. After this, the doctor waves the dung round the patient, and afterwards when it is examined it is found to contain queer, hairy creatures. "People marvel at this and are convinced that the man had been bewitched, and that the funny-looking creatures came out of his body. For this the patient is made to pay a heifer. And this for nothing more than a conjuring trick since these creatures could not possibly exist in a man's body."

To contain the dung while he ceremoniously waves it over the afflicted area, the "trickster" uses a small cloth or leather pouch attached to a chain of brightly colored beads. You guessed it! Although in black and white, the photograph accompanying the warning was of a necklace and pouch identical to the one brought home from Maasailand.

The Exorcism

At 8:45 p.m., as I walked up the sidewalk toward the front door, I wavered; I reminded myself why I was doing this. If self-generated healing is possible--and we know it is--then self-generated harm is, too. At bottom, this is no different than any of my other cases. Whether she's doing it consciously or not, this woman is cutting herself. There's no other explanation. All I have to do is put a block in that process.

I joined them--Lauren, Rochelle, and Chicky–in the living room. I had brought the candles and incense, the Gregorian chants on a cassette tape, and a heavy crucifix that hung lopsidedly from a wide red ribbon. A vial of Father McNeal's holy water sat on the mantel close by. The pets had been shooed outside to insure their safety. We chatted quietly. The mood was appropriately somber. In the shadows, the house had all the charm of a funeral parlor. Lauren and I sat on wrought iron chairs: mine at a right angle to hers. On doctor's orders I was to keep an eye on her scar-patterned arm. Rochelle and Chicky were at a table ten to twelve feet in front of us. I lit the candles and the incense burner; put the tape on, and seized the vial of holy water. Then, piously, I picked up the oversized crucifix and, flicking water in our direction, moved to put it around Lauren's neck. Moving back to the cluster of candles, I sorted through the pages of my ritual and began the ceremony with authority in my voice.

The formal rite of exorcism, the seventeenth-century *Rituale Romanum, is* forty pages of Latin, and applies only to direct possession by a demonic power. Pope Leo XIII's *Prayer Against Satan and The Rebellious Angels is* translated and can be used by a lay person in a simple ceremony. There's a shorter three-page rite for diabolical infestation that requires no special church permission and is written in English. I drew heavily from them all. I improvised. I cut here, combined there, and came up with a prayer of expulsion worthy of The *Malleus Maleficarum.*

What right did I have to plagiarize these solemn exercises? One theory claims that a lay-person may be more successful performing them than a priest. The demon, so the theory goes, is quick to leave his victim when there's no opportunity for him to posses an official member of the church. I was about as *unofficial* as you can get. I don't put much stock in demons. But, no matter the cause of her predicament, my pleas were meant to be helpful--not heretical.

I spoke now in a soft murmur--in monotone, word for word. Occasionally, underlined phrases got special emphasis; but always it was one unvaried tone. The candles and threw an eerie light across Lauren's face. The girls, nearly out of their flickering reach, were motionless--in semi-trance. They watched their mother like onlookers at an inquisition. My eyes were fixed on her thin-fleshed arm, uncovered for the ceremony. If she was going to slash herself, I'd be the first to know. I had been disinclined--and with good reason--to leave it unprotected; she'd been cut so often. But her confidence in me was essential to the mission. If the "expert" had to rely on magic symbols wouldn't they risk losing that confidence?

I droned on. My thoughts wandered to her over indulgences. How could the events of the past six weeks be pinned on tobacco and alcohol abuse? It made no sense. How could it be fraud? What could she gain by deceiving me? Anyway, didn't the severity of the wounds eliminate that possibility?

After an hour, the chants merged into one mesmerizing sound. The incense burned my nose and fogged my brain. Lauren's face gave away nothing in the reflected light. If the job was getting done you'd never know it by her expression. She just sat there, staring beyond her daughters--the weight of the crucifix bending her forward slightly.

Now the climax! My deep, hail and brimstone pronouncements rang from the cathedral ceiling: "I cast thee *out,* thou unclean spirit, along with the least encroachment of the wicked enemy, and every phantom and diabolical legion!" With that, Lauren sat up stiffly and arched her back. Her arms fell to her sides. Her upper body was rigid, eyes bulging from their sockets. She kept gazing over the heads of her girls. I was tempted to look--to see what it was that held her transfixed; but I didn't. I was determined not to be distracted from my watch. No matter, I didn't have long to think about it. While still seated, she and her chair slid backwards about six feet. Then, before I could react to prevent it, she came flying out with a force that catapulted her to the foot of the girl's table: a trip of some fifteen feet!

All I could think was *My God, I've killed her! How could I have talked myself into this?* Meanwhile, I went on without missing a beat. Lauren had virtually soared to her daughters, who sat there frozen to their chairs. I helped her back to hers, then stood behind it--one hand on her forehead, the other on her shoulder. It was a precaution well-taken. In a few seconds, her body stiffened, shook slightly and pressed forward. It was all I could do to restrain her. The exorcism ended at five minutes past midnight. My bedeviled client went limp. Rochelle rushed to her side. For the first time since the ordeal had begun Lauren allowed herself the luxury of tears. When they stopped, I asked why she'd pushed away from me:

Lauren: "I didn't do it. I didn't push myself...I was carried away. The chair *glided* back. Look here ... do you see the feet on these chairs? They're levelers (metal disks). You can't move them on a shag carpet ... they get all hung up." She was right. Even without her in it, the chair was so entangled in the loops of the rug it wouldn't budge.

Andrew: "Are you hurt? Did you jump ..... leap out of it?"

Lauren: "I have no memory of it; none whatsoever. Oh! my hip (gently rubbing it)." "I remember seeing the ghost from my bedroom. It was up against that wall [gesturing behind Chicky, still seated at the table]. After I got back in my chair, it changed into a vision of St. Michael, the Archangel. "It's over, isn't it? I've had, what you might call, an *emotional climax.* But now it is over. *My problem is over!"*

Andrew: "That's wonderful. How do you know that?"

Lauren: "*He* told me. I cannot say how, but he told me. He said, 'All is well.' He was no longer the threatening vision of my dreams. It turns out, you see, that his aim was never to hurt me. He was coming to protect me from evil. His sword was there to protect not harm me."

Andrew: "Well, then, that's it, isn't it? It's really over."

Later that morning, Lauren left a message for me to call her. I had mixed feelings about it, I confess. It may sound callous, but I hoped I'd never hear from her again. The saga of the Kanes had been like a novel with a happy ending. I liked the way it came out, but enough was enough! I didn't want it to start up again--to hear anymore for fear that it *hadn't* ended. And then what? What would I do for an encore? But I had to call her back. I couldn't just ignore her.

According to bible tradition, Friday the thirteenth hasn't been what you'd call a “whoopty-do” day in human history. It's the day Eve was tempted by the serpent; the day the Flood carried off all but Noah and his passengers; it's when Solomon's Temple was destroyed; and, of course, Jesus of Nazareth was crucified on that unluckiest of days. For a fifty-five-year-old Florida divorcee, Friday, July 13, 1989, marked the end of a bloody confrontation with the unknown. Lauren Kane awoke that morning feeling like Ebenezer Scrooge on Christmas Day--full of the spirit of life:

Lauren: “I had the best night's sleep in months. I am emotionally drained. But, for the first time since whatever it was started, 1 slept the whole night through. No dreams. No ghosts. And I owe it all to you.”

Thank Goodness I sighed to myself.

Lauren: “Best of all, I've thrown my crutch ... the bandages away! I even threw away the paper I had the symbols drawn on. I do not need them. I've had not one attack since last night: that's almost thirteen hours.”

Andrew: “What about Dr. Pauling. You *are* going to keep seeing him?”

Lauren: “*For what purpose* (she growled)?”

Andrew: “Now, Lauren. I can hear the waves of euphoria sweeping over you, and I'm tickled. I'm sure you won't have any more problems (with my fingers, toes, and eyes crossed), but you've been through quite a trying experience. Don't you think he shared in your cure?”

Lauren: “I do not!”

One of the things I learned while researching this case was that after a successful expulsion the individual relieved of the oppressing influence must be guided toward a new lifestyle. This can't mean a return to a previous way of life, if that involved habits that may have helped to create the sense of obsession. If my ritual created a special kind of emotional crises, one that enabled Lauren's old behavior patterns to breakdown, then new, constructive patterns needed to be created. The only person I trusted to direct her along that path was Dr. Pauling.

Andrew: “I believe you're cured, Lauren. I don't want to overdramatize this thing, but I hate to think what awaits you in the future if you are unwilling to make changes in your life now.” My words were ignored. She never went back to him. On August 29, 1979, after a month-plus of being phenomenon free, there was a setback. While dining with her daughter Rose, Lauren lost the sight in her left eye.

She went immediately to Orlando Regional Hospital where an ophthalmologist ordered tests, then advised her to consult a psychiatrist:

Rose: "It's ironic, really. My mother was instrumental in forming the first organization to help pre-school sightless children under the auspices of the American Society for the Blind. And here she was blind herself. Thank goodness it only lasted a day." "How does a person suddenly loose sight in one eye, then get it back in less than twenty-four hours?" "Central serous chorioretinopathy, a form of detached retina that can be precipitated by highly stressful situations, is one way," I said. "It's a blood pressure problem," I went on, "but doesn't automatically correct itself. Then there's psychological trauma ... hysterical blindness. When that happens there's nothing physically wrong with the eyes. The patient says he can't see, but there's no pathology. Basically, she's mimicking it. It's not that difficult to tell if she is, though."

I called the department of ophthalmology at Orlando Hospital. "Why," I wanted to know, "had someone recommended a psychiatrist to Mrs. Kane?" They refused to disclose any specific information to me, as I was reasonably certain they would. Still, the doctor on staff volunteered that temporary loss of sight could be due to psychological trauma. The first step in diagnosing the condition was to eliminate injury and disease as factors. If they were absent, brain waves would be measured to see if the patient was imitating sightlessness. In the test, a bright light is repeatedly flashed until it evokes electrical waves from the visual cortex. The blind do not register responses to the light. One who is unconsciously faking it has no conscious awareness of the light, but his brain gives him away by registering a normal response.

Later, Lauren confirmed that a painful light, "like a psychedelic strobe," had been flashed in her eyes during the test. She had been unable to detect it, she said, when her right eye was covered. Apparently, her brain registered otherwise. Hysteria is an overwhelming invasion of the conscious mind by the unconscious. It's a complex neurosis that takes a number of forms. One of the most important varieties is "conversion hysteria," in which mental conflicts are converted into physical symptoms: functional disorders such as paralysis, deafness and blindness, and may--as we shall see--trigger spontaneous psychosomatic bleeding. (Source: *Dictionary Of Psychology,* J.P. Chaplin, Dell.)

An important characteristic of hysteria is hyper suggestibility. As far back as 1926, researcher Harry Price investigated a suspected Poltergeist involving an impressionable Romanian girl named Eleonora Zugun. After being the center of psychic activity for several months, the tragic young woman was thrown into a lunatic asylum. When at last she was released, the "spirit" tormenting her began slapping her face. It tossed her out of bed, filled her shoes with water, and scratched her forearm mercilessly. Even when the area involved was covered, wounds would appear without a visible external source. Finally, it became bored with the girl and left. Author Colin Wilson *(Poltergeist: A Study In Destructive Haunting,* Perigee Books, 1981), not one to attribute these forces to subconscious externalizations, believed the girl's grandmother--by suggesting that she was in league with the devil--had been responsible for triggering the attacks. Harry Price had come to the same conclusion. Price was certain the scratches, in particular, were a form of self-punishment brought on by guilt, likening them to stigmata.

In a literal sense, Lauren Kane was a stigmatic. No, she didn't suffer lacerations on the body in imitation of the wounds of Christ, but she was a "sufferer" by anyone's definition. In his *Handbook Of Parapsychology,* A.R.G. Owen tells us there is a parallel between haunted people and stigmatics, many of whom show overt hysteria:

> *"However, it is easy to recognize those factors of personality which are common to the generation of hysteria and the adoption of stigmata. These are a kind of autosuggestion, and a capacity for unconscious control of bodily processes manifested in skin lesions and blood flow in the stigmatic (the bleeding cuts), and in conversion symptoms in the hysteric (the loss of sight)."*

There are those who contend that such attacks are not uncanny; that the resulting gashes, punctures, and marks of a scourging do not come from out of the blue. D.H. Rawcliffe, in *Occult and Supernatural Phenomena* (Dover Publications, New York) called them "neurotic excoriations": scratches and strips of skin missing from various areas of the body. The most common is dermatitus artefacta, or self-inflicted lesions. Interestingly, they nearly always appear in more or less symmetrical form.

In spite of the witnesses who saw her arm open on its own (including yours truly), we cannot, unequivocally, rule out Lauren's physical participation in them. With hysterical individuals, a deliberate infliction of wounds could easily develop a delusion that they were done supernaturally. I was not persuaded that they were supernatural or paranormal events, only that they were abnormal. What I'm driving at is, Lauren Kane's neurotic delusion--if, indeed, that's what it was--could well have become her observer's optical illusion. What we thought we saw was not necessarily what occurred. Actually, once we had ruled out *deliberate* fraud, it really didn't matter much.

Random Adolescent RSPK mimics the actions of a spoiled child in the midst of a tantrum (slamming doors, throwing windows open, ringing doorbells, etc.). Cases that begin with scratching sounds are fairly typical of the Poltergeist, too. Is it possible these effects were taken over by a malicious spirit? Or is Mrs. Kane a latent psychic sensitive? Some of the phenomena that have revolved around her seem to indicate that she is. An English medium, Mrs. Fielding, a subject of study by famed psychic researcher and psychologist, Nandor Fodor, was thrown out of her chair by an unseen force on many occasions. Fodor tells us that after succumbing to the emotional pressures of her gifts, Mrs. Fielding began going hysterically blind, and insisted she'd been clawed by an invisible tiger.

I feel relatively safe in saying that evil spirits, invisible tigers, and miraculous religious intervention played no part in the stigmata of Lauren Kane, or in her brief loss of sight. In the final analysis, Lauren, like the unfortunate Romanian girl, had a deep-seated reason for self-punishment: an emotional conflict alluded to by Dr. Pauling early in his treatment, yet unknown to me. In Max Longs, *The Secret Science Behind Miracles,* the author speculates that a good deal of what we call madness is a kind of haunting by spirits; with only those suffering from deep-seated guilt feelings likely to be affected by it. But what was this lady guilty of? And what form of madness is it that drives a person to thrust bloody wounds upon themselves? To end it all is one thing! To bring on so much unnecessary misery is quite another.

Ultimately, she acknowledged her grief. In October 1988, nearly eight months before the first attack on her arm, Lauren's mother had passed away. They'd been extremely close, this diminutive, hard-boiled woman and her equally formidable mom. After the fire, Lauren's great joy over the return of the finger ring *("It was given to me by my mother! after all. ")* was some indication of the depth of her feelings for the aging woman:

Lauren: "When no one else could stand us, *we* could stand each other. She was my buddy ... always at my house, or we'd be going places together. She celebrated her eightieth birthday on Columbus day, 1988 ... enjoyed better health than most sixty-year-olds, except for one annoying ailment: a painful hammer toe. "I must say, her constant complaining about it ..... the way it would rub itself raw against the top of her shoes, threw me! Why in the world did she put up with it? 'Get it fixed,' I told her,' '*It's no big deal. "* Not adverse to surgery herself, Lauren had talked her into the operation:

Lauren: "For her birthday, I treated. She probably could've bought the hospital ... but I took the excuse that it cost too much away from her." "Well, the operation was a success. But, as the saying goes ... the patient died. She was out of recovery, in her own room when of all things she fell out of bed. Someone had forgotten to put the railing up. My mother, a woman who hardly knew a sick day in her life, died of a blood clot. (Her voice unsteady now.) She broke her hip and a clot worked its way up to her lungs, then to her heart. I was devastated (the last word barely audible)."

This is a woman not given to exaggerated sentimentality. She knew perfectly well that she was not responsible for the fate of her mother. But, then, we all have a better handle on our conscious anxieties than those hidden deep inside.

So what's the verdict? It would appear that *The Exorcist* connection, the *Hairballs from Hell,* and the burden of her mother's death all came together in a bizarre scenario that inspired the 1989 round of phenomena. What caused the two previous rounds was no longer important. Was I justified in performing an exorcism? Would I have done so had we known of Lauren's self-accusations--the hidden guilt? Modern medicine, psychology, and parapsychology make it impossible to believe that demonic obsession exists. Yet a properly structured exorcism, like the Psi Session, its counterpart, has a beneficial psychotherapeutic function.

It may be that she empathized with the young priest in *The Exorcist:* "Why you do this to me, Demi?" his mother had said, as she neared death in a strange hospital room. It was a guilt-laden scene that could have touched off the attacks that followed. Lauren may have atoned for the death of her own mother--the debt having been fully paid when the Archangel conveyed the message that all was well.

Thanks to "the scourge of demons," she was free from physical attack with only one minor exception. At three o'clock on the morning of June 25, 1990, nearly a year after the exorcism, I got a ship-to-shore phone call from a disgruntled voyager: "I'm on a cruise off the coast of Maui," she shouted. "As I was preparing for dinner this evening, my arm opened. Just a tiny scratch, but it did bleed for a length of time. What shall I do? I can't very well put those grotesque symbols on ... not with all the social events aboard ship. We dined at the *Captain's* table tonight." Groping for words, I paused for a moment, then answered, "I know it's embarrassing, Lauren, but what else can I offer you?"

I got back in bed and quietly protested: *St. Michael, where are you when we need* you? I was almost asleep when Father Bowdern's diary came floating by my eyelids. I knew I was up to stay--being one of those poor souls who, once they're up can never get back to sleep--so I looked for my official copy of the Jesuit's tale. *Let's see. What was it he wrote about the boy and the scourge of demons? Oh, yeah. Here it is:*

> *"On the night of April 18, 1949, the ritual resumed. Father Bowdern forced Roland to wear a chain of religious medals and to hold a crucifix in his hands .... Roland exploded. Five Alexian brothers rushed to hold him while he screamed he was one of the fallen angels...the boy brushed off his handlers and soared through the air at Father Bowdern standing at some distance from his bed. "Bowdern persisted in the ritual, reciting incessantly until 11:00 p.m. when Roland interrupted. In a completely new masculine voice (he) said, 'Satan! Satan! I am St. Michael! I command you, Satan, and the other evil spirits to leave this body, in the name of Dominus, immediately! Now! Now! Now!' "'He is gone!' (the boy) said wonderfully, with a smile on his face. He told the priest he had had a vision of St. Michael holding a flaming sword. He felt wonderful."*

Chapter 12: When Common Sense Falls Victim to Uncommon Events

"The mind of man is capable of anything--because everything is in it, all the past as well as all the future."
(Heart of Darkness, Joseph Conrad)

I remind the reader, as I so often remind my clients, that ghost hunting is much more an art than a science. The results parapsychologists achieve are open to many different interpretations, a few of which find us divided in our thinking.

I have a healthy skepticism of the supernatural. I believe that most, if not all explanations lie in the realm of the paranormal: a force in our physical world whose underlying principles, although not yet discovered, are discoverable. Psychic ability is purely a natural thing and a part of all of us. We humans leave some sort of energy, a psychic impression on our surroundings, particularly when we're experiencing trauma. Those who have developed psychic abilities (clairvoyance, precognition, or telepathy) come along, 'see' the imprint and call it a ghost or spirit. When people see these strange impressions they don't realize they're coming from their own mind; that they don't exist anywhere *but* the mind.

I remain open to other possibilities but am nearly convinced that when apparitions and ghosts are observed, and where it appears that messages are coming from the spirit world, the activity is originating within the mind of one of the residents of the troubled house. And the attendant effects?: They're a creation of the witnesses' subconscious, activated by the impression (only) of a haunting entity. Ghosts exist, but there's every reason to suppose they're a subjective creation generated in, and projected from the brain; and very little to indicate they represent the activities of a surviving personality. The culprits in my cases are generally the haunted themselves.

The majority of my case histories (as well as those gathered by other researchers) confirm that ghosts are subjective creations. A living person is the genesis, and the unnerving spectacle, which seems to fly in the face of reason, is created by mental and physical pressures, or by nervous system dysfunction.

It is tempting to attribute all hauntings to the hidden secrets of human and physical nature, The Psychic Imprint, what researcher H.H. Price called "place memories," is a case in point. For if events are somehow enrolled in Nature's logbook and later triggered to play back their scenes and sounds, then a major portion of the enigma has been solved. All of them could be attributed to a kind of trace recording left behind; and any house wherein men have lived and died might very well be haunted--haunted by "place memories."

Imprints, however, merely suggest that a replica or image of the body survives death in the same sense that we all survive in our photographs and voice recordings. They can't explain how apparitions of the dead are seen in places they did not visit, let alone live or die in; they can't account for the evidential information that ghosts bring back with them.

By definition, Imprints, like motion pictures, exist in the past and are unable to bridge the gap to the present. (If we could speak to movies, ten-thousand moonstruck women would have saved Patrick Swayze from becoming a *Ghost*). In a number of cases, acts are carried out on behalf of the "visitor" and the haunting terminates abruptly: a less dramatic yet equally successful method of exorcism. Imprints seem never to end, often enduring as long as the house in which they appear--even longer in some cases. Although impressive, in my mind, Psychic Imprints and Random and Directed RSPK provide only a convenient solution to the problem, not a comprehensive one. No matter how hard I try I can never seem to squeeze all of my cases into one or more of these categories. On rare occasions--as, I believe, has been demonstrated in this work--the information received appears to have come from someone other than our clients.

I'm not ready to pronounce this outside source indisputably "out of this world": I'm 99.9% sure it isn't. On the other hand, I have always subscribed to the philosopher's tool known as "Occam's Razor": "When two explanations of a phenomenon are available, one complex and the other simple--prefer the simple."

I've heard some of the most incredibly complex theories advanced over the years to explain-away phenomena--phenomena, which, in truth, could more easily be explained as supernatural. The report that came from a woman in Marietta, Pennsylvania led to a classic example of this tendency to invent ultra-rational explanations for simple, straightforward occurrences.
The woman, whose privacy I will respect, wrote:

> *"Several years ago, my mother-in-law was stricken critically ill and passed away at the hospital. Later, as we stood by the bedside, the nurse apologized and said she could not find her lower dentures (which, incidentally, my mother-in-law had never worn because they were too uncomfortable). I didn't pay much attention for I felt it would make little difference. "Now the arrangements for burial had to be made and once back home my husband decided he would take the deed to the cemetery lot to the undertaker. While he was gone, I poured a cup of coffee, lit a cigarette and tried to relax. I was alone in the kitchen of our two story house. Suddenly, I heard a knocking sound. I couldn't imagine where it was coming from. At first, I paid no attention. My mind was simply too full of the events of the last few days. But the knocking persisted and eventually I became aware that this was not a constant noise but sounded only now and then. Now I was determined to find out where the noise was coming from. I might add that I did not, at any time, liken this noise to the recent death. I did not feel anything other than the knocking noise was annoying."*

It took my correspondent two additional, full hand-written pages to detail all the places she searched before finding the source of the intermittent rapping: the attic.

> *"I went directly to the door, slid back the latch and started up the stairs. As I did, the thought occurred to me that if anything were loose and banging the sound would not have been so loud to me two stories below. Our attic is finished and divided into two rooms. Why I did not bother with the front room, I d not know. I went directly to the back room, directly to a shelf above the antiquated water tank, directly to the small suitcase on that shelf. Without hesitation I flipped the latches open and lifted the lid. There, in a corner was a small box. I immediately lifted it out and opened it. What did I find? My mother-in-law's lower dentures! "My first act was to call the undertaker and tell him I had found the dentures and I was sure my mother-in-law would look better in repose with them in. He agreed. (She had such a weak chin.) My second act was to listen for any knocking sounds. There weren't any. Not that day or since then."*

After reading this account, I asked friends and associates--those whose ideas on the subject ranged from slightly credulous to fanatical disbelief--to tell me what they thought was behind the case of the "clacking choppers." Here are some of the responses: "The whole incident was a coincidence ... pure and simple." "Originally, the woman knew where the dentures were, then forgot, then remembered again ... all tricks of the memory." "She's a clairvoyant or a psychometrist..." After she touched the uppers, it was no problem for her psychic powers to find the lowers." "Her husband knew subconsciously where his mother's teeth were and telepathically transmitted the location, by way of the rapping sounds." "She saw it in a dream and subconsciously *created* the noise in order to locate it more dramatically." "The story is a hoax. Why would her mother-in-law want to be buried in uncomfortable teeth, anyway?" "The Lord works in mysterious ways."

I concede that just because a case appears to be supernatural doesn't mean it is. What bothers me is the fact that those who advance convoluted ideas to explain-away the things that happen are not open-minded enough to even admit the possibility exists. Most parapsychologists agree on the part the subconscious mind plays in hauntings, except for one detail. Some researchers believe that entities and their energy source, without exception, *originate* in and are *directed* by the living. In cases of Random and Directed RSPK, and in a few Psychic Imprints, I agree. As for those leftover--the ones that defy classification--I'm still not convinced. Not that it is incumbent upon me to know Nature's secrets; but it is a fascinating question, isn't it?

It's true that a case for the supernatural has not yet been made. I have never observed anything to *convince* me that there is or isn't an afterlife or a spirit plane. All the so-called evidence is circumstantial. Scientific logic insists on a natural explanation. I normally find one. Such reasoning rejects the existence of an initiating

Agent--living or dead--and places responsibility for the disturbance squarely on the imagination and suggestibility of those who report ghostly experiences.

As I enter my second quarter-century of trying to "bust the ghosts," I've come to realize that common sense sometimes falls victim to uncommon events. Though occasionally given to mysticism, when circumstances seem to support them, I prefer scientific explanations for what I come up against. I've always believed that these traits were a source of strength and ultimately an advantage to my clients, the folks who must face these things, whatever they are. A skeptical friend once told me, "If you believed in everything you'd drop off the deep end." I say, if I believed in nothing I'd volunteer to leap off. In the world of parapsychology, a Ghost Detective must operate somewhere in between.

THE END

About the Author

Dr. Andrew Nichols is a psychologist, parapsychologist, and investigator of hauntings, poltergeist cases and other paranormal phenomena. He is a member of the American Psychological Association, the Parapsychological Association, and the Society for Psychological Hypnosis. He has been a psychology professor for many years and is adjunct faculty (parapsychology) at Santa Fe College (Gainesville, FL). Nichols is listed in Who's Who in American Teachers.

During his 30 year career, Nichols has investigated more than 600 reported cases of ghosts, hauntings and poltergeists, and conducted studies in telepathy, precognition, and paranormal dream experiences. He has written numerous articles on paranormal experiences for popular magazines such as Fate and his scientific papers on the paranormal have been published in scientific journals such as the Journal of Parapsychology, International Journal of Parapsychology, European Journal of Parapsychology, and Proceedings of the Parapsychological Association.

Professor Nichols has presented lectures and workshops on paranormal topics at colleges and conferences throughout the U.S., Canada and Europe. His work has been featured in many books on paranormal topics.

As a media consultant on paranormal topics, Nichols has appeared on numerous TV and Radio programs in the U.S., Canada, Europe and Japan, including Unsolved Mysteries; 48 Hours; Inside America's Courts, and was a recurring guest on NBC's The Other Side. Several television specials have featured Dr. Nichols' work, including ABC's The World's Scariest Ghosts, The Discovery Channel's Real Ghosthunters and Ghost Detectives, and Beyond Death on A&E.

He has investigated alleged poltergeist disturbances for government agencies and law enforcement, including The United States Army, Oak Ridge National Laboratory and the Daytona Beach Police Department. In 1999 he was co-recipient of a grant to study haunting and poltergeist cases, the first grant of its kind in the history of psychic research.